Current Topics in Microbiology 226 and Immunology

Editors

Springer
Berlin
Heidelberg
New York
Barcelona
Budapest
Hong Kong
London
Milan
Paris
Santa Clara
Singapore
Tokyo

DNA Vaccination/ Genetic Vaccination

Edited by H. Koprowski and D.B. Weiner

With 31 Figures and 14 Tables

Springer

Hilary Koprowski, M.D.

Professor
Thomas Jefferson University
Department of Microbiology and Immunology
Center for Neurobiology
Rm. M-85, Jefferson Alumni Hall
1020 Locust Street
Philadelphia, PA 19197-6799
USA

David B. Weiner, Ph.D.

Associate Professor
University of Pennsylvania School of Medicine
Hospital of the University of Pennsylvania
Department of Pathology and Laboratory Medicine
500 Stellar-Chance Bldg.
422 Curie Boulevard
Philadelphia, PA 19104-6100
USA

Cover Illustration: Plasmid delivery to a host tissue results in tissue specific protein production and specific activation of T cells, B cells and antigen presenting cells generating specific immunity (by courtesy of Bin Wang and Michael Chattergoon).

Cover Design: design & production GmbH, Heidelberg

ISBN-13: 978-3-642-80477-9 e-ISBN-13: 978-3-642-80475-5
DOI: 10.1007/978-3-642-80475-5

Library of Congress Catalog Card Number 15-12910

Softcover reprint of the hardcover 1st edition 1998

Typesetting: Scientific Publishing Services (P) Ltd, Madras

SPIN: 10503929 27/3020/SPS – 5 4 3 2 1 0 – Printed on acid-free paper

Preface

Genetic / DNA immunization represents a novel approach to vaccine and immune therapeutic development. The direct injection of nucleic acid expression cassettes into a living host results in a limited number of its cells becoming factories for production of the introduced gene products. This host-inappropriate gene expression has important immunological consequences, resulting in the specific immune activation of the host against the gene-delivered antigen. The recent demonstration by a number of laboratories that the induced immune responses are functional in experimental models against both specific infectious diseases and cancers is likely to have dramatic consequences for the development of a new generation of experimental vaccines and immune therapies. This technology has the potential to enable the production of vaccines and immune-based therapies that are not only effective immunologically but are accessible to the entire world (rather than just to the most developed nations).

Vaccine Development

Vaccination against pathogenic microorganisms represents one of the most important advances in the history of medicine. Vaccines, including those against polio, measles, mumps, rubella, hepatitis A, hepatitis B, pertussis and other diseases, have dramatically improved and protected more human lives than any other avenue of modern medicine. The vaccine against smallpox, for example, has been so successful that it is now widely believed that this malicious killer, responsible for more deaths in the twentieth century than World Wars I and II combined, has been removed from the face of the earth. Traditional vaccination has relied on two specific types of microbiological preparations for producing material suitable for immunization and the generation of a protective immune response. These two broad categories involve either live infectious material which has been manufactured in a weakened or attenuated state, thus preventing the vaccine from inducing disease, or alternatively, nonlive, inactivated, or subunit preparations. Live attenuated vaccines, such as the polio and smallpox vaccines, for example, stimulate protective immune responses as they replicate in the host. However they are limited in their ability to replicate in humans and, accordingly, do not normally induce crippling or lethal disease in inoculated individuals. Since these live preparations are actually produced in the host as the attenuated virus replicates within the host cells, they have several unique immunological properties associated with this in vivo replication. The viral proteins produced within the host are leaked or perhaps shed into the extracellular space surrounding the infected cells and are then picked up, internalized, and digested by scavenger cells that patrol the body. These include both macrophages and the several forms of dendritic cells, as well as B cells; collectively they function to expand immune response. These antigen presenting cells (APCs) then recirculate small fragments of the antigen to their surface, attached to MHC II or class II antigens. This complex of foreign peptide antigen plus host MHC class II antigens forms part of the specific signal with which APCs themselves, along with the MHC peptide complex, trigger the action of a central population of immune cells, the T helper lymphocytes cells. The second part of the activation signal comes from the APCs themselves: they display on their cell surface, in addition to MHC-antigen complexes, costimulatory molecules. Together they can drive T cell expansion and activation through interaction with their respective ligands, the T cell receptor complex (TCR) and the costimulatory receptors CD28/CTLA4, present on the T cell surface. Activated T helper cells secrete soluble molecules that serve as powerful activators of immune cells. In addition, as viral proteins are produced within cells of the host, small fragments of viral proteins are drawn to the cell surface, chaperoned by the host cell MHC I antigens. These complexes are recognized by a second class of T cells, the killer or cytotoxic T cells. This recognition, coupled with secondary stimulation provided by professional APCs and helped by cytokine production from the stimulated helper T cells, either individually or collectively, is responsible for the development of the mature effector T cytotoxic cells capable of destroying viral factories, i.e., infected cells. In addition, in most instances live infection induces an additional benefit, that of lifelong immunity.

In contrast, killed or nonlive vaccines composed of whole or even fragments of viruses, when inoculated as vaccines, are not produced within host cells and upon administration mostly wind up in the extracellular space and not within cells. Accordingly they provide protection by directly generating T helper and humoral immune responses against the pathogenic immunogen and are therefore particularly useful in protecting against infection, especially when antibodies are responsible for microbe inactivation. In the absence of the cellular production of the foreign antigen, these vaccines are usually devoid of the ability to induce significant T cytotoxic responses. The lymphokine profiles of these antigens are also often different from those induced by live attenuated vaccines. In addition these vaccines are not actually produced in the host and therefore are not customized by the host. The immunity induced by these vaccines frequently wanes during the lifespan of the inoculated hosts and may often require repeated immunizations or boosting to achieve lifelong immunity. However, nonlive vaccines offer some important advantages over live vaccines. They are generally easier to manipulate and usually easier to produce than live vaccines. Furthermore they enjoy an inherent advantage: we can design them to contain only the specific antigenic target of the pathogen that is involved in the development of protective immunity and exclude all other viral components. The latter, namely, can deflect the focus of the immune response and actually have unwanted immunological cross-reactivities that can cause problems for the vaccinated host.

Accordingly, live attenuated preparations are clearly the vaccines of choice in terms of the diverse immune responses they produce. However they pose risks of reversion to a more pathogenic form: during their replication process they can sometimes convert back into an actual disease-causing virus. Furthermore they may retain the unwanted characteristic, even in the attenuated state, of inducing disease in persons with weak or compromised immune systems. This includes persons suffering from genetic or drug induced immune deficiency, receiving cancer chemotherapy, or suffering from a separate infection that weakens their immunity (AIDS patients, the elderly). An individual vaccinated with a live attenuated virus can pass that attenuated virus on to others, thus presenting risks of inadvertent spread to an unknowing host (something health care workers need to keep in mind). Mothers have become vaccinated with the live attenuated polio vaccine simply by changing the diapers of their recently vaccinated infants. In contrast, safely manufactured nonlive vaccines that have specific immunity-inducing abilities have no risk of inducing infection or disease due to a need to maintain a replicating phenotype. They can be designed to exclude those portions of a pathogen that may induce unwanted cross-reactive immune responses. Accordingly they have been the vaccines of choice, based on both their safety and specificity and certain manufacturing considerations.

Recent work from a number of laboratories (including that of the editors) has demonstrated that inoculation of a DNA plasmid directly into a host with subsequent expression of the encoded peptide sequences in vivo results in presentation of the specific encoded protein(s) to the immune system. Since DNA vaccines are nonreplicating and are produced within the host cells they can be constructed to function with the safety advantages of a subunit nonlive vaccine and yet to mimic immune potentiating aspects of a live attenuated vaccine. This represents a distinct advantage for initial vaccine and analysis as DNA vaccination represents a rapid system for directly testing subunit vaccination strategies without viral production or protein purification procedures. The latter can significantly slow down the investigation process. Furthermore, DNA vaccination facilitates a diverse array of immune analysis from a single immunization platform. It may be particularly advantageous that such a system can deliver an antigen which could be presented for development of both the helper T cell and the cytotoxic T cell arms of the immune system in the absence of live vector construction. Using direct DNA immunization the genes cloned into the expression vectors can be manipulated to present single proteins or an extended genome with only the genes that might lead to pathogenesis removed. Accordingly, this approach, if translatable into success in the clinic, may present the best of both prior vaccine modalitieswith the added benefit of decreased turnaround times for vaccine development.

History of DNA Inoculation in Vivo

The ability of genetic material to deliver genes for therapeutic purposes has been appreciated for some time. Some of the earliest experiments describing the transfer of DNA into the cells of a living animal were reported by Stasmey (1950), Parchkis (1955), and Ito (1958). These reports described the ability of chromatin preparations, crude preparations of DNA isolated from tumors, to induce formation of tumors after injection in laboratory rodents. Later experiments further purified the genetic material and confirmed that direct DNA gene injection in the absence of viral vectors can result in the expression of the inoculated genes in the host. If genes encoding cancer antigens were injected, cellular transformation and cancer, with a reproducible frequency, would result in the inoculated animals. Studies using purified DNA and RNA derived from viruses were also reported. Most importantly, Atanasia and colleagues (1962) demonstrated that the subcutaneous administration of purified polyoma virus DNA transforming sequences as part of the viral genome to newborn hamsters resulted in the generation of anti-polyoma antibodies to the virus and the induction of tumors from the transforming genes. In a similar study, Orth and coworkers (1964) injected either newborn or 40-day-old hamsters with DNA isolated from polyoma virus grown in tissue culture. Thirty days following subcutaneous injection of purified DNA, 34 of the 35 newborn hamsters had developed antibodies and 26 developed tumors. The results described by these studies are explained by the uptake of the injected DNA by cells of the inoculated host followed by translation of the injected DNA sequences into messenger RNA. The host translation system then drives expression of viral encoded genes and their functions, which in the case of the administration of transforming gene sequences, is cellular transformation. Crucial for this discussion however is the outcome of this delivery of viral genes: it resulted in the presentation of these genes to the immune system, leading to the production of an immune response against the virus, as determined by the in vivo antibody production observed in these studies.

Additional experiments extend these findings to recombinant DNA molecules. Israel and colleagues (1979) reported that recombinant clones of purified DNA containing head to tail dimers of the polyoma virus resulted in seroconversion of inoculated animals after injection of such DNA vectors into weanling mice. Similar experiments using plasmids containing hepatitis B genetic material were reported by Will (1984). The in vivo gene expression of plasmid material was further studied in regard to successful insulin delivery as a gene therapy approach in living animals (Dobensky 1985). Elegant studies from the laboratory of Jon Wolff (University of Wisconsin) detailed the long-term expression of injected plasmids in vivo (1990). Together, these studies and others served to support the idea that purified nucleic acids could be directly delivered into a host and proteins would be produced. Such proteins could be immunogenic in some settings. However, challenge studies in these systems have not yet been widely reported.

In 1992 Tang and Johnston reported that the delivery of human growth hormone gene in an expression cassette in vivo resulted in protein expression in the inoculated animals. These authors utilized a genetic gun to shoot gold particles coated with DNA through the skin layers of mice. The inoculated animals produced detectable levels of human growth hormone. Interestingly they reported that antibodies developed against the human growth hormone produced in the mice. These authors termed this immunization procedure genetic immunization, describing the ability of inoculated genes to be individual immunogens. These studies elegantly brought together and focused the earlier work in this area.

Anti-Pathogen Immune Responses Induced By DNA Immunization In Vivo

Almost simultaneously with the publication by Tang and Johnston the annual vaccines meeting at Cold Spring Harbor in the fall of 1992 brought together several independent groups that collectively reported on the use of DNA immunization technology for the generation of protective immune responses in vivo. These presentations, which used unique approaches and were directed at different viral targets, were not the first public presentations of any of the laboratories involved. However, the collective success of these independent investigators and the large amount of data they presented from diverse systems could not be ignored by the vaccine community. Specifically, Margaret Liu of Merck and her collaborators at Vical reported on the induction of immune responses to influenza A virus following intramuscular inoculation of mice with highly purified plasmid DNA encoding influenza A genes. Following three inoculations over 6 weeks using 100g doses of plasmid DNA, the resulting immune responses to the influenza proteins included antibodies, cellular immune responses, and evidence of protection from viral challenge. Harriet Robinson and colleagues (University of Massachussetts) utilized the gene gun to deliver influenza genes in plasmids in vivo; they observed that DNA doses on the nanogram level were able to induce both antibodies and cellular immune responses. These doses were also able to protect mice from challenge in a similar influenza challenge model. One of us (Weiner) reported that direct injection of plasmids encoding core HIV genes, rev and the envelope genes induced humoral and cellular immunity against HIV-1, the virus which causes AIDS. The immune responses induced were able to prevent HIV infection in in vitro assays.

There have been many important studies, which, together, form the basis of this emerging field. Some of the seminal pioneers in this arena have contributed chapters to this volume. Specifically, using a variety of delivery methods and DNA constructs, DNA immunization in animals has been reported to generate host immunity against herpes simplex virus (B. Rouse and C. Pachuk), human hepatitis B virus (H.L. Davis), human hepatitis C surface virus (J. Wands), HTLV-I (K. Ugen), HIV-1 (M. Bagarazzi, S. Lu, and J. Warner), malaria (S. Hoffman), cancer-related antigens and oncogenes (R.C. Kennedy and W. Williams), and veterinary targets (L. Babiuk).

Collectively the studies presented here serve to highlight several important features of DNA immunization. Several different tissues, but especially those of muscle and skin, can serve as antigen factories, serving to induce immune responses in vivo. These studies suggest that one important issue is the nature of the antigen presenting cell involved in generating the DNA-based responses. Protective cellular and humoral responses can be developed through a variety of delivery methods and these appear to be long lived. Very low levels of antigen are necessary for this antigenic stimulation; in fact, using the gene gun, nanogram levels appear functional in mice. Facilitated delivery through chemical reagents such as charged lipids, local anesthetics, and gold particles can have positive consequences for immunization and in vivo gene delivery, and they can significantly improve in vivo gene delivery.

The properties reported for DNA immunization warrant significant attention. Nanogram levels of expression vector are sufficient for production of antibody responses using gene gun technology, while microgram levels of expression vector are similarly sufficient in direct intramuscular or intradermal injection studies. Chemical agents increase the potency and lower the amount of DNA necessary to achieve specific immune responses in vivo: DNA immunization induces detectable seroconversion in mice in as little as 12 to 21 days following intramuscular injection of only 5–10g of antigen encoding plasmid. Multiple plasmids encoding several different antigens can be immunized against in a single administration. Examination of the duration of seroconversion in mice demonstrates that humoral immune responses can persist for more than 1 year, which represents up to half of this species' lifetime. Similar studies in primatcs support a shorter half life of immune responses induced in vivo. Both T cell proliferation and cytotoxic T-lymphocyte (CTL) responses can be induced in a single immunization. The CTL responses can be boosted by repeat immunizations without vector interference, and this holds for both rodent and primate studies. The effects of boosting on the resulting humoral responses are not as clear, at least not in mice, and are under investigation in primates, but they seem supportive of a boosting requirement. Importantly though, DNA immunization can be boosted by reboosting with purified protein antigen or native antigen encoding the same antigen as is present in the expression cassette. This process demonstrates two important features of DNA immunization: (a) it presents antigens which are cross-reactive with native antigen and (b) significant immunological memory can be induced. In mice at least DNA immunization tends to favor a Th1 vs Th2 type of response; this is based on antibody profiles as well as on cytokine induction and the cellular nature of the induced response.

Clinical Studies of DNA Vaccines

Recently, clinical trials of DNA vaccines have commenced. Currently, clinical studies are being pursued on a variety of infectious agents including HIV-1, HBV, influenza, HSV, and malaria. Studies are also progressing on the treatment of human cancer, including colon carcinoma and lymphoma. Specific safety concerns include monitoring for anti-DNA immune responses and for any pathology at the sites of injection. Preliminary reports suggest that these vaccines have been well tolerated in humans. This will no doubt encourage further examination of this technology and allow us to develop it to its full potential. Furthermore, there are encouraging signs from the informal reports of the immune responses observed in the first group of patients studied. This is an extraordinary development. In only a few years this technology has moved from the bench to the clinic. Assuming the safety profiles of these vaccines hold up, the flexibility of the technology will no doubt lead to the evaluation of its potential for controlling many of the world's most terrible infectious killers.

While the future of this technology appears rosy, its actual appearance is likely to change radically as we attempt to modify, improve, and combine this technology with other important approaches. As the first gene therapy to have ever been tested on normal healthy humans its application seems unlimited. By building on the combined results of studies such as those presented in this volume we will likely be able to ride this technology well into the future. It is important to remember though that other technologies have enjoyed equal early appeal. Specifically, recombinant protein vaccine, peptide vaccines, and live recombinant viral vaccines (such as the poxviral vaccines) also received early rave reviews. Yet more than a decade after their appearance only a single representative of this group of high technology approaches, the recombinant hepatitis B vaccine, has been successful in the clinic. This sobering thought suggests that while there are no quick fixes, it is the concerted and constant attention to the development of a new field that is necessary for its ultimate success. Using the studies collected in this volume as a starting point, we look forward to a long and productive journey along this newest path of vaccine development.

Philadelphia,
November 1997

Hilary Koprowski
David B. Weiner

List of Contents

List of Contributors

(Their addresses can be found at the beginning of their respective chapters.)

DNA Vaccine Strategies for the Treatment of Cancer

P.A. BENTON and R.C. KENNEDY

1 Introduction

A wide variety of immunotherapeutic strategies for the treatment of cancers is currently under investigation in both animal models and in clinical human trials (reviewed in DRANOFF and MULLIGAN 1995). These include in vitro expansion and adoptive transfer of tumor-specific cytotoxic T cells in the presence or absence of recombinant cytokines (ROSENBERG et al. 1985, 1986, 1988, 1989, 1991, 1993; TENG et al. 1991), tumor antigen-based vaccines (CHEEVER et al. 1986; BRIGHT et al. 1994a; CONRY et al. 1994; STEVENSON et al. 1995), and tumor-reactive monoclonal antibodies conjugated to toxins (TRAIL et al. 1993; VIVETTA et al. 1993). One of the most recent and innovative strategies has been based on the concept of genetic immunization. It is now well established that vaccination with plasmid DNA

Department of Microbiology and Immunology, University of Oklahoma Health Sciences Center, 940 Stanton L. Young Blvd, BMSB 1053, Oklahoma City, Oklahoma 73104, USA encoding a specific gene can elicit strong, long-lived protective immune responses to a variety of infectious agents (reviewed in ULMER et al. 1996; WHALEN 1996) as well as certain tumors (CONRY et al. 1994; NABEL et al. 1994b; STEVENSON et al. 1995; BRIGHT et al. 1996; CORR et al. 1996; SCHIRMBECK et al. 1996). DNA vaccines directed at eliciting an immune response toward tumors appear to offer promise for both the prophylactic and therapeutic treatment of cancer.

In this review, we will discuss the application of DNA and genetic immunization to the field of cancer therapy. This will include the description of direct approaches that target tumor-specific or tumor-associated antigens and indirect approaches that result in the expression of molecules that activate the immune system in general. This review will encompass the discussion of these approaches as they have been applied to several different animal models, as well as preliminary human clinical trials.

2 Strategies for Gene Delivery

Several techniques have been employed for the delivery of nucleic acid vaccines to animals and humans. These methods have a variety of advantages and disadvantages, but overall many have been successful in the stimulation of specific immune responses. A major goal of these gene delivery strategies is to enhance the in vivo transfection efficiency and the subsequent expression of the encoded gene product. Below, we will briefly describe strategies for gene delivery that have been employed both in vivo and in vitro.

2.1 Delivery Vectors

To date, a number of studies have examined several different parameters of in vivo gene transfer. These include DNA structure, route of injection, promoters, and injection vehicles. Most investigators have found that simple saline solutions often represent reasonable carriers of injected DNA when an intramuscular route of immunization is employed (ULMER et al. 1993). DNA prepared by methods that typically enhance in vitro transfection (calcium phosphate precipitates and/or liposomes) may or may not enhance the efficiency of in vivo gene transfer. Several reports have suggested that co-administration of toxic agents, such as the anesthetic bupivicaine and cardiotoxin, along with DNA can enhance gene transfer and expression in vivo (DAVIS et al. 1995). Presumably, administration of the toxic agents resulted in skeletal muscle damage and subsequent repair events may have led to a non-specific increase in gene transfer in this dual delivery system.

The type of plasmid and its design appear to influence the successfulness of efficient gene expression and the induction of an immune response. Most studies have utilized the human cytomegalovirus (HCMV) immediate early promoter in the plasmid constructs to enhance gene expression at the injection site (FOECKING and HOFSTETTER 1986). Other promoters have also been tested for their ability to improve the efficiency of gene expression following in vivo DNA transfer, although there is not as much information regarding these. Although the objective is to generate levels of gene expression sufficient for the induction of long-lasting immunity, the long-term effects of expression from these highly efficient promoters in human skeletal or dermal tissue are unknown.

2.2 Nontumor Targeted DNA Vaccination

Most genetic vaccination research to date has involved the injection of DNA vaccines by either subcutaneous, intradermal, or intramuscular routes (WOLFF et al. 1990; ULMER et al. 1993; RAZ et al. 1994). These methods depend on the ability of surrounding cells to uptake the injected DNA that is subsequently transcribed and translated into the protein against which an immune response is generated. Another means of gene delivery is via injection of DNA-coated gold beads by particle bombardment (gene gun). Due to the force of the injection, this manner of gene delivery ensures that the DNA enters the cells, and therefore does not rely upon DNA uptake by the host cell itself. Although not as thoroughly investigated, a few reports studying the effects of intravenous injection of DNA have also demonstrated the ability to detect the expression of injected genes in many organs (ZHU et al. 1993; reviewed in HASSETT and WHITTON 1996). All of these routes of DNA delivery have been shown to be effective in the elicitation of protective immune responses toward infectious agents as well as demonstrating responses involved in tumor rejection. Relevant to the treatment of existing tumors via DNA injection, however, are strategies that specifically target the DNA preparation to the tumor cell.

2.3 Tumor Targeted DNA Vaccination

In addition to the routes and parameters of gene delivery just mentioned, DNA vaccines can be directly targeted to the tumor site by several different mechanisms to ensure gene expression at the tumor site. These means increase the probability that the tumor cells alone will be targeted for destruction, eliminating the likelihood that healthy bystander cells will be killed.

One method of directly targeting vaccines is through intratumor inoculation. Direct injection of DNA into the tumor guarantees that expression of the protein will be from within the tumor. This method has been utilized in the induction of tumor immunity against CT26 colon adenocarcinoma (H-$2K^d$) transplantable tumors that are normally poorly immunogenic (see Sect. 3.1). Briefly, allogeneic MHC class I antigen genes are injected into these tumors, resulting in an enhanced recognition of the tumors by the immune system, and decreases in tumor size. This approach provides an indirect means for improving the antigenicity of an existing tumor and also demonstrates that in certain cases DNA vaccination may offer therapeutic protection even after tumors have been identified.

Another means of directly targeting DNA vaccines to the tumor site is to place the gene of interest under the control of a tissue-specific promoter, so that gene expression occurs only at the tumor site. This technique has been illustrated in at least one model system in which the 5′ flanking region of the tyrosinase or tyrosinase protein 1 (TRP-1) was shown to be adequate to direct the expression of β-galactosidase in human and murine melanoma cell lines and melanocytes (Vile and Hart 1993). Notably, expression was not detectable in a wide range of other cell types, thus demonstrating the restricted specificity of gene expression. Gene expression was also enhanced by an additional step of direct intratumor DNA injection. These experiments illustrate the possibility of directly targeting specific genes to the tumor site if adequate information is available regarding tumor-specific promoters.

Receptor-mediated delivery of DNA is a third possible method by which genes can be targeted directly to tumors. This method has been tested primarily in non-tumor models, but has the potential for application in cancers when tumor-specific surface markers are known. These experiments target the DNA by the attachment of specific proteins that interact with cell surface receptors. Studies using this technique have mainly employed adenovirus particles or influenza fusogenic peptides for mediating gene delivery, since the viral attachment proteins will specifically interact with cell surface receptors (Wagner et al. 1992a, b; Cristiano et al. 1993a, b). Although this technique has not yet been utilized in targeting tumors, it has great potential for delivering DNA directly to tumors via tumor surface markers.

3 Mechanisms of Tumor Rejection

Several methods have been attempted to stimulate protective immune responses against tumors in vivo. These methods generally fall into two different categories: (1) indirect approaches that involve the enhancement of a generalized immune response, or (2) direct approaches that result in the induction of a specific response to a tumor-associated or tumor-specific antigen. Although the focus of these categories is different, both types of approaches have proven to be effective in the development of protective immune responses. Several examples of these techniques will be discussed below.

3.1 Enhancement of Immune Responses by Indirect Approaches

Immune responses are initiated and enhanced in response to many different stimuli. Examples include the presence of foreign antigens, the release of cytokines and/or lymphokines, and the presence of costimulatory molecules and cell adhesion molecules, to name a few. In cases in which no information is available regarding tumor antigens or tumor surface markers, indirect or nonspecific approaches for the enhancement of an anti-tumor immune response have been designed and tested. These methods include the introduction of MHC class I genes into tumors, the increased expression of costimulatory molecules, and the increased production of cytokines, all of which can result in the stimulation of local or systemic inflammatory responses.

It has been observed that many solid tumors, leukemias, and lymphomas have decreased surface expression of MHC class I molecules (reviewed in BROWNING and BODMER 1992; MOLLER and HAMMERLING 1992). The mechanisms by which some tumors down-regulate the expression of MHC class I molecules include the loss of peptide transporter genes, mutations in the β2-microglobulin gene, and dysregulation of transcription (ROTH et al. 1994). Since tumors frequently display low numbers and/or the absence of MHC class I molecules, it has been postulated that lack of tumor antigen presentation results in the escape of tumors from immunologic surveillance. These hypotheses are supported by experiments in which transfection of tumors with MHC class I genes resulted in the enhancement of an immune response toward these tumors and tumor rejection (HUI et al. 1984; TANAKA et al. 1985; WALLICH et al. 1985). In these cases it appeared that increased surface expression of MHC class I molecules on tumor cells subsequently increased tumor antigen presentation, thus stimulating a protective anti-tumor immune response.

In a different approach, another indirect means of stimulating an immune response at the tumor site has been demonstrated by the transfection of allogeneic MHC class I genes into tumors. Plautz and colleagues have demonstrated the ability to stimulate an immune response by the transfer of allogeneic MHC class I genes into the poorly immunogenic CT26 adenocarcinoma tumors of mice (PLAUTZ et al. 1993). These experiments demonstrated that intratumor injection of a recombinant H-2K^{s} gene into CT26 tumors in BALB/c (H-2K^{d}) mice induced the generation of splenic lymphocytes with lytic activity against CT26 targets (PLAUTZ et al. 1993; NABEL et al. 1994b). In addition, in mice presensitized with tumor cells expressing H-2K^{s}, tumor size decreased following vaccination with H-2K^{s}-encoding DNA (PLAUTZ et al. 1993). An important observation in this study was the ability to activate tumor-specific immune responses by the expression of the foreign MHC class I molecules on the adenocarcinoma tumors that were normally poorly immunogenic. These experiments demonstrated the ability to boost the immune response by altering the context in which tumor antigen is presented to the immune system, and thus provide a good model for studying the effects of direct DNA inoculation into preexisting tumors for the stimulation of a therapeutic effect.

Stimulation of strong immune responses also requires specific costimulatory signals in addition to the initial T cell receptor (TCR)/MHC interactions. Normally, antigen presenting cells deliver this signal through interactions of costimulatory molecules on their surface and that of the target cell. The interaction of CD28 on the T cell with B7 on the antigen presenting cell appears to be essential for appropriate T cell activation and proliferation. Since most tumors are derived from parenchymal or mesenchymal cells that do not express B7 it has been hypothesized that this may be another reason, along with decreased MHC class I expression, that tumor cells are poorly immunogenic (ROTH et al. 1994). The introduction of the B7 gene into existing tumors has indeed been shown to enhance an anti-tumor response. Studies employing the poorly immunogenic, B7 negative, murine melanoma cell line K1735 have demonstrated that transfection of the B7 gene into these cells and subsequent tumor challenge into naive, syngeneic mice led to tumor rejection (CHEN et al. 1992; TOWNSEND and ALLISON 1993). It has been suggested that the presence of B7 permitted the proper costimulatory signals for interleukin (IL)-2 production and subsequent proliferation of tumor-specific $CD8^+$ cytotoxic T lymphocytes (CTLs) (SCHWARTZ 1992).

The potential anti-tumor effects of many different immunomodulatory molecules, including IL-1, IL-2, IL-3, IL-4, IL-6, IL-7, IL-12, interferon-γ (IFN-γ), tumor necrosis factor (TNF), transforming growth factor-β (TGF-β), granulocyte colony stimulating factor (G-CSF), macrophage colony stimulating factor (M-CSF), and granulocyte-macrophage-colony stimulating factor (GM-CSF) have been analyzed in many different systems. In all cases, except with M-CSF, GM-CSF and TGF-β, the expression of these molecules led to the rejection of transfected tumors (reviewed in DRANOFF and MULLIGAN 1995). The basis for the priming effects that cytokines impart on DNA vaccination came from the work of Ertl and colleagues in the murine rabies virus system (XIANG and ERTL 1995). These investigators demonstrated that co-injection of a DNA plasmid encoding genes from rabies virus glycoproteins with naked DNA encoding the murine GM-CSF enhanced both antibody- and cell-mediated responses to the rabies virus glycoprotein.

In a recent study, cytokines have been shown to enhance DNA immunization resulting in the effective treatment of established pulmonary metastases in mice (IRVINE et al. 1996). In this system, the murine CT26 tumor was modified to express β-galactosidase (β-gal) as a tumor-associated antigen. Pulmonary metastases were established in mice employing this modified tumor and it was demonstrated that DNA immunization with a plasmid expressing the immunodominant β-gal peptide had little to no impact on the growth of the established metastases. However, when cytokines were incorporated as adjuvants for the DNA immunization, a significant reduction in pulmonary metastases was observed. Also effective in enhancing the tumor therapy in this model was human IL-2, mouse IL-6, human IL-7 and mouse IL-12. The most profound effect was demonstrated with mouse IL-12. These findings clearly indicate that cytokines involved in the activation and expansion of lymphocyte populations improve the therapeutic effects of DNA vaccination. Indeed the implementation of cytokines as adjuvants for DNA vaccination represents an indirect means for activating specific immunologic responses that may play an important role in tumor immunity.

In several other studies it has been shown that the direct injection of tumors with the genes coding for immunomodulatory cytokine molecules induces a local and sustained inflammatory anti-tumor response (TEPPER et al. 1989; WATANABE et al. 1989; Fearon et al. 1990; Gansbacher et al. 1990a, b). In these cases, increased expression of specific cytokines through the genetic modification of tumors led to both a decrease (or abrogation) of tumorigenicity and an increase in anti-tumor immune responses. The increased expression of these immunomodulatory molecules can enhance CTL activity against tumor cells and thereby decrease tumor size. Clinical trials have been performed in which advanced malignant melanoma patients have been treated with autologous tumor infiltrating lymphocytes (TILs) transfected with the TNF gene or the IL-2 gene (Rosenberg et al. 1992a, b). Genetic modification of TILs (transfected with either IL-2 or TNF genes) greatly reduced the risk for toxic side effects that have been demonstrated in the systemic administration of IL-2 and TNF.

The data presented above emphasize the importance and potent effect that immunomodulatory molecules have in directing a protective immune response toward tumors. These novel approaches to increase the expression of costimulatory molecules, cytokines and interferons should enhance our understanding of tumor immunity and provide practical applications for the therapeutic or prophylactic treatment of cancers.

3.2 Induction of Tumor-Specific Immune Responses by Direct Approaches

Evidence for the existence of tumor-specific antigens was initially determined by experiments in murine model systems in which tumor-specific immunity was induced by vaccination with inactivated parental tumor cells (Gross 1943; Foley 1953; Prehn and Main 1957). In these studies, mice immunized with chemically induced tumor cells were able to reject grafts of the same tumor, and these responses were shown to be tumor-specific. These studies provided support for the search for tumor-specific antigens in human tumors, in the expectation of finding targets for immunotherapy. Unfortunately very few tumor-specific antigens have been identified in human cancers. Certain tumor-associated antigens characterized by their preferential expression on tumors, however, have been identified. These tumor-associated antigens have been employed as targets for the induction of tumor-specific immune responses.

Anti-tumor immune responses can be generated directly toward the tumor when there is information regarding the presence of tumor-associated antigens, tumor surface markers, or tumor-specific proteins. As mentioned above, the site of gene expression can be controlled by the inclusion of tumor-specific promoter regions within the plasmid DNA, or by the conjugation of the DNA with proteins that interact with tumor-specific surface receptors. These strategies result in the expression of the injected gene, such as a toxin or cytokine, only at the tumor target, thus ensuring tumor-specific destruction or a localized immune response, respectively.

Since plasmacytomas and B-cell lymphomas overexpress surface immunoglobulin (Ig), these molecules have been targeted as tumor-specific antigens.

Therefore, studies in murine models have been designed to test the effects of vaccination with idiotypic Ig protein and idiotypic DNA vaccines. The first successful demonstration of induction of anti-tumor immunity was shown in a murine plasmacytoma model (EISEN et al. 1975). Anti-idiotypic antibodies (anti-Id) were generated in response to vaccination of mice with idiotypic protein, and the immunized mice were protected from subsequent tumor challenge.

The murine lymphoma BCL1 represents another model being utilized for the characterization of tumor-specific antigens and the development of immunotherapy strategies. In B cell lymphomas the surface immunoglobulin is the tumor-specific surface marker that can be targeted in vaccines, and DNA vaccines have been produced which induce anti-idiotypic antibodies. This system will be discussed in detail (see Sect. 4.2).

It is well known that simian virus 40 (SV40) causes the transformation of tissue culture cells and induces tumor formation in mice. Many studies have shown that the immunization of mice with syngeneic inactivated SV40-transformed cells or SV40 large T antigen (T-ag) results in anti-tumor protection when mice are subsequently challenged with viable tumor cells (reviewed in TEVETHIA et al. 1980; SCHIRMBECK et al. 1992; BRIGHT et al. 1994a). In SV40 murine tumor models, intramuscular, intraperitoneal or subcutaneous immunization of BALB/c mice with plasmid DNA encoding the SV40 T-ag elicits a protective anti-tumor immune response (BRIGHT et al. 1996; SCHIRMBECK et al. 1996). These studies will also be discussed in detail (see Sect. 4.3).

Due to the limited information regarding tumor-specific antigens, the direct approach of targeting genetic vaccines toward specific tumors is more difficult than inducing a generalized immune response at the tumor site. However, in cases in which tumor antigens have been identified, tumors have been targeted successfully using DNA immunization technologies, resulting in the induction of protective immunity.

4 Studies That Examine DNA Based Cancer Vaccine Strategies in Animal Models

A variety of studies employing animal tumor models have demonstrated the importance of T cells in the regression of tumors. Adoptive transfer experiments demonstrated that a specific subset of T cells, TILs, are potent effector cells involved in tumor regression. Studies from animal models have demonstrated that TILs are more effective in mediating in vivo regression of tumors than are IL-2 lymphokine activated killer (LAK) cells. The identification of tumor-specific or tumor-associated antigens that are recognized by T cells may be invaluable in the development of anti-cancer vaccines. One of the major advantages of DNA vaccines is the ability to induce cell-mediated immune responses, including the production of MHC class I-restricted CTLs. Thus, DNA vaccination strategies have the potential to induce antigen-specific T cell responses that can target tumor cells expressing these antigens. Below, we will briefly describe the limited number of studies employing the application of DNA vaccines to target tumor cells in a direct manner.

4.1 Melanomas

Metastatic melanoma still represents one of the most difficult tumors to treat. The reported objective response rate of melanoma to chemotherapy is less than 25% and the cure rate is virtually 0% (CARREL and RIMOLDI 1993). Fortunately, the cure rate of early melanoma results in a much better prognosis. Accurate early diagnosis of precursor lesions and of primary tumors is critical. The characterization of melanoma antigens that may represent targets for immunotherapy has relied primarily on molecular biology techniques for their identification. In general, the cloned melanoma antigens fall into three categories (reviewed in KAWAKAMI et al. 1996). The first category includes a number of antigens that represent putative tumor-associated antigens because they are expressed in the retina and uvea as well as in melanocytes. Tyrosinase, gp100, MART-1/Melan A, and TRP-1 appear to be present in the melanosomes of pigmented cells. Tyrosinase and TRP-1 represent enzymes that are involved in the synthesis of melanin. The second category of antigens includes members of the MAGE gene family and the recently identified melanoma antigen termed BAGE. These antigens appear to be expressed only on adult testis among normal individuals and are also expressed in a number of other tumor cells, including breast, lung, and bladder cell carcinomas. The first two categories represent tumor-associated antigens and may be considered as cancer regression antigens. The final category is reserved for a group of tumor-specific antigens. Included are epitopes expressed on melanoma cells that result from point mutations, deletions, recombination events, and errors in transcription of the normal gene products. These mutated forms of tumor-specific antigens tend to be expressed on only a small number of the tumors and may not be good targets for immunotherapy. Additionally, oncofetal proteins that are not expressed on adults tissues represent potential tumor-specific antigens. Examples of this category of melanoma antigens includes β-catenin and p15. Direct application of DNA vaccination strategies includes immunization with plasmid DNA encoding cancer regression antigens or tumor-specific antigens. Alternatively, one could express on the melanoma cell an antigen that would be considered foreign by the host and employ a tumor-specific immunotherapeutic approach by targeting the expressed foreign antigen.

Murine models of melanoma tumor formation and treatment have been developed and DNA vaccines have been designed to specifically target melanoma cells. Targeting of these vaccines to melanoma tumors was accomplished in one study by direct intratumor injection of a DNA construct encoding the β-galactosidase gene under the control of a melanocyte-specific promoter (VILE and HART 1993). This vaccine was able to stimulate β-galactosidase expression in B16 melanomas in syngeneic mice, but not in Colo 26 tumors. These experiments displayed the ability to directly and specifically stimulate the expression of genes at the tumor site and offer a strategy for limiting gene expression to the tumor cells. This technology could result in the development of a vaccine which encodes a toxic protein that will only be expressed at the tumor site, thereby eliminating the possibility that healthy cells will be destroyed, as in traditional chemotherapy regimens.

In another murine tumor model, B16 brain melanoma cells were genetically engineered to produce one of the following cytokines: IL-2, IL-3, IL-4, IL-6, IFN-γ, or GM-CSF (Samson et al. 1996). This study was performed to determine whether cellular vaccines expressing specific cytokines could affect tumors at distal immunologically privileged sites. When C57/BL/6J mice were vaccinated with untransfected or transfected B16 cells and challenged one week later with untransfected cells, the mice receiving GM-CSF-producing B16 cells had statistically longer survival times than all other animals. The other cytokines did not appear to have any statistically noticeable effect upon survival time in this model. The most notable finding in these experiments was that vaccinations by cytokine-producing cells could affect an immunologically privileged site, such as the central nervous system. These experiments, therefore, have important implications in the treatment of established brain tumors by this type of immunotherapy.

4.2 Idiotypic DNA Vaccines Against B-Cell Lymphomas

As noted above, B cell clones resulting from neoplastic transformation in vivo express idiotypic determinants that can act as tumor-associated antigens. In this context, DNA vaccines have been developed which encode variable region sequences that can induce the formation of anti-Id (Hawkins et al. 1993, 1994). An idiotypic DNA plasmid vaccine has been developed which encodes the scFv sequence from the murine BCL1 lymphoma surface Ig under the control of the Rous sarcoma virus long terminal repeat (RSV LTR) (Hawkins et al. 1993). Intramuscular injection of this plasmid induced the production of anti-Id in approximately 50% of the mice and this antibody reacted against BCL1 tumor cells (Hawkins et al. 1993). Since results were variable, the addition of cytokine-encoding DNA plasmids was incorporated into the vaccine strategies (Stevenson et al. 1995). Additional constructs were made that encoded either IL-2 or GM-CSF in order to test their effect when co-injected along with the idiotype vaccine plasmid. Interestingly, intramuscular injection of both DNA vaccines led to an increase in anti-Id, compared to administration of the idiotypic vaccine alone. Unfortunately, the vaccinated mice did not survive challenge with BCL1 tumor cells. It was hypothesized that a block in antibody activity may have resulted due to the formation of immune complexes consisting of the secreted idiotypic scFv antigen from the injected plasmid and the induced anti-idiotypic antibodies (Stevenson et al. 1995).

These studies demonstrate the effectiveness of DNA vaccines in eliciting an immune response against specific tumor-associated antigens. This lymphoma model is one of a few tumor models available for the analysis of DNA immunization against tumor-specific antigens. Small preliminary clinical trials have begun in human lymphoma patients who do not respond to conventional chemotherapy (STEVENSON et al. 1995).

4.3 SV40 Murine Tumor Model

Recent studies have provided direct evidence that SV40 may play a role in a variety of human brain tumors and malignant pleural mesotheliomas (KRIEG et al. 1981; reviewed in PASS et al. 1996).Therefore, an SV40 murine tumor model provides a useful tool in the development of tumor vaccine strategies. SV40 is a polyomavirus that grows in rhesus and cynomolgus monkey kidney cells without causing any cytopathogenic effects. Along with SV40, two human polyomaviruses have been isolated and described. These viruses, designated JC and BK, share nucleotide homology with SV40. The genome of these viruses encodes an early region for the large and small tumor antigens and a late region that encodes the viral capsid and noncoding regulatory region. SV40 can cause a lytic or abortive infection in cells depending on whether the host cell type is permissive or nonpermissive to SV40 replication. The SV40 proteins associated with in vivo oncogenesis and in vitro cell transformation includes the early region-encoded tumor antigens. The SV40 T-ag has the capacity to bind the cellular tumor suppressor gene products p53 and retinoblastoma protein (Rb). SV40 T-ag is also involved in the initiation of viral replication in permissive hosts and regulates the expression of the late structural gene products that associated to form virions.

Although mice represent a nonpermissive host for SV40 replication, inoculation of SV40-transformed mouse kidney fibroblasts, designated mKSA cells, into syngeneic mice results in lethal tumor formation. A number of investigators have demonstrated that SV40 T-ag-specific immune responses are capable of providing protective tumor immunity in murine systems (reviewed in PASS et al. 1996 and TEVETHIA et al. 1980). In early studies, immunization of BALB/c mice with SV40 T-ag, either purified from transformed cell lines or expressed on inactivated syngeneic transformed cells, resulted in protection from a subsequent lethal tumor challenge (ANDERSON et al. 1977; CHANG et al. 1979). Mechanism(s) of this tumor immunity did not appear to involve CTL responses as BALB/c mice (H-2d) generated little to no detectable SV40 T-ag-specific CTL activity following immunization with purified SV40 T-ag or SV40 T-ag expressing syngeneic cells (GOODING 1977; KNOWLES et al. 1979). The mechanism(s) of tumor immunity in this system remained undetermined for a number of years, however, it was clear that SV40 T-ag was a tumor-specific antigen that was responsible for protective tumor immunity.

Our laboratory began evaluating mechanism(s) of tumor immunity specific for SV40 T-ag in murine systems following immunization and/or challenge with SV40 transformed cells in 1985 (KENNEDY et al. 1985, 1987). Since that time, we have evaluated a number of prophylactic and therapeutic vaccination strategies within the SV40 T-ag system. These strategies include idiotype-anti-Id (Kennedy et al. 1985; Mernaugh et al. 1992; Shearer et al. 1993), SV40 T-ag synthetic peptides (Bright et al. 1994b), recombinant SV40 T-ag (Bright et al. 1994a), and nucleic acid (Bright et al. 1994a, 1996). Although each of these four strategies demonstrated some degree of tumor immunity within BALB/c mice, the recombinant SV40 T-ag and DNA vaccination strategies were superior and resulted in complete immunity in most instances. These studies demonstrated a dichotomy in the mechanism(s) of tumor immunity within the murine SV40 T-ag system and clearly suggested an individual role for both humoral and cell-mediated immune responses in protection of BALB/c mice from a subsequent challenge with a lethal tumor burden.

In 1994, our laboratory described the ability to induce SV40 T-ag-specific CTL responses in BALB/c mice by injection of plasmid DNA encoding SV40 T-ag (Bright et al. 1994a). This result was surprising in light of previous studies that characterized BALB/c mice as nonresponders for CTL induction by SV40 T-ag (Gooding 1977; Knowles et al. 1979). Apparently, delivery of SV40 T-ag via plasmid DNA was capable of rendering the non-CTL responding inbred strain of mice into a responding strain. This was the first description of the induction of SV40 T-ag-specific CTL responses in $H\text{-}2^d$ strains of mice. In addition, this was one of the first studies to demonstrate that DNA immunization was capable of generating immune responses to a viral-encoded tumor-specific antigen. This suggested the potential for genetic immunization as a direct approach for cancer vaccination strategies. In subsequent studies, we evaluated the protective capacity of this DNA vaccination strategy. Intramuscular vaccination of mice with a DNA plasmid encoding the SV40 T-ag prior to challenge with mKSA cells led to protection against tumor formation and prevented death from a lethal tumor burden (Bright et al. 1996). Serum from immunized mice failed to exhibit significant levels of SV40 T-ag-specific antibodies, however, in vitro CTL assays showed that DNA vaccination induced MHC class I-restricted CTLs that specifically lysed SV40 transformed cells (Bright et al. 1996). DNA vaccination against SV40 T-ag, therefore, resulted in the induction of SV40 T-ag-specific CTLs which were protective against lethal tumor challenge. These results were recently confirmed by other investigators (Schirmbeck et al. 1996). Since the induction of CTLs appears to play an important role in tumor rejection, this SV40 murine tumor model provides a very beneficial system for the evaluation of the protective correlates of tumor immunity induced by genetic vaccination.

4.4 Carcinoembryonic Antigen Vaccines

Carcinoembryonic antigen (CEA) is one of the most well characterized human tumor-associated antigens. CEA and α-fetoprotein, categorized as oncofetal proteins, were the first tumor-associated antigens to be identified and are used as tumor markers in the diagnosis of certain tumors. The CEA molecule is a glycoprotein of approximately 200 000 kDa, heterogenous in nature, and appears to be a member of the Ig supergene family of proteins (ZIMMERMAN and THOMPSON 1990). It is found in the alimentary tract, liver, and pancreas of fetuses from 2 to 6 months gestation. These proteins do not completely disappear in adulthood and are detectable by sensitive immunohistochemical methods in normal adult tissues, such as the colon, and in the plasma. CEA is produced in high concentrations on certain types of tumor cells and is present on the surface of adenocarcinoma cells in a wide variety of human cancers. In addition to increased levels of CEA being present in certain malignancies, CEA can also be elevated in nonmalignant diseases, such as chronic obstructive pulmonary disease and cirrhosis. So, contrary to early expectations, these tumor-associated antigens are not tumor-specific. However, since they are expressed at much higher levels in patients with tumors, detection of these antigens has proven helpful in diagnosis and therapy, and CEA can serve as a target for immunologically based cancer therapies.

A human CEA-expressing syngeneic murine colon carcinoma cell line exists and is being utilized in the development of a tumor model for the investigation of vaccines directed against CEA. In one study, a DNA plasmid encoding the full length human CEA cDNA was placed under control of the cytomegalovirus (CMV) early promoter/enhancer for use as a genetic vaccine (CONRY et al. 1994). Intramuscular injection of this plasmid into C57/BL/6 mice induced humoral and/or cellular immunity specific for human CEA in all immunized mice (CONRY et al. 1994). Anti-CEA antibody was detectable in four out of five mice, but not in mice receiving control vector plasmid. CEA-specific memory T cells were detected in three of five mice by lymphoproliferation assay, indicating a specific immune response had been induced by DNA vaccination. These studies did not include any tumor challenge data, but do demonstrate the ability of DNA vaccination to elicit a CEA-specific immune response and may be applicable to therapeutic regimens in the future. Since CEA is expressed by adenocarcinoma cells in so many different cancers, this new genetic vaccine model provides a useful tool for the development of vaccine strategies of several different cancers.

4.5 Other Tumor Model Systems

Additional investigators have employed DNA immunization to evaluate its potential as an immunotherapeutic strategy against cancer. In one study, investigators employed a surface marker on human lymphocytes as a tumor-associated antigen by expressing this molecule on a murine tumor cell line. Specifically, the human CD4 molecule was expressed on the cell surface of the murine myeloma cell line SP2/0 and was utilized as a tumor challenge model in mice. In this study, the human CD4 molecule represented the tumor antigen target for immunotherapeutic intervention. Immunization with plasmid DNA encoding the human CD4 gene protected mice from a lethal tumor challenge with CD4-expressing SP2/0 myeloma cells (WANG et al. 1995). Characterization of the immune response resulting from vaccination with this plasmid DNA demonstrated both humoral and cell-mediated components. CTL responses were demonstrated by the observation that immune lymphocytes were capable of lysing syngeneic target cells expressing the CD4 molecule.

Another recent study has employed the use of mutated self or altered self nucleotide sequences to induce specific CTL responses (CIERNIK et al. 1996). The relevance of this approach to tumor immunity would be in cases in which the mutated self protein represents an epitope on a tumor suppressor gene product, such as p53 or Rb. In this scenario, the majority of self protein present on both normal and tumor tissues would be unaltered. Therefore, immunization with the entire protein could induce immune responses against the unmutated regions of the molecule and effect its normal function. In these cases, DNA immunization strategies could be employed to induce an epitope-specific immunity that targets only the mutated region, or mutated form of the self protein, ensuring that normal self proteins would not be affected.

5 Human Trials

With the discovery of tumor-associated antigens, came the expectation that human cancer vaccines could be developed. Indeed, many cancer vaccines have been developed and they primarily incorporate either whole antigen or peptides in a variety of adjuvants or genetically engineered virus vectors (reviewed in COHEN 1993). Most of these vaccines have targeted melanomas and colon cancers, some utilizing purified antigen and others using crude vaccines containing a mixture of potential tumor antigens. The nature and design of the vaccine play a significant role in its ability to induce protective tumor immunity. The design of these cancer vaccines was aimed at either the stimulation of a humoral antibody response or a cell-mediated T cell response. It still remains to be determined which arm of the immune system most effectively eliminates malignancies, but it is clear from several studies that both responses can lead to tumor rejection.

Since genetic vaccination is a relatively new field, few human trials have been initiated involving the direct inoculation of DNA vaccines into human cancer patients. In addition, due to potential safety concerns, genetic immunizations have only been carried out in patients in advanced stages and whose tumors have been refractory to traditional cancer therapy regimens.

Studies by Rosenberg have pioneered the field of gene therapy. Early clinical trials demonstrated that ex vivo incubation of cancer patients lymphocytes in IL-2 activated a population of lymphocytes which could kill natural killer (NK) cell-resistant tumor cells, but not normal cells (MULE et al. 1984). These cells were termed LAK cells or large granular lymphocytes demonstrating LAK activity. Further studies included the infusion of IL-2 along with the LAK cells finding that IL-2 had several toxic side effects in these patients. However, 13%–57% of patients with advanced cancer displayed partial or complete regression, depending on the tumor type (reviewed in ROSENBERG 1991). Patients with renal cell carcinoma, non-

Hodgkin's lymphoma, melanoma and colorectal cancers had the best results. These trials were followed by ex vivo expansion of T lymphocytes isolated directly from patient tumors (TILs) and re-infusion into patients. In the case of melanomas, about 38% of patients showed at least a partial response (ROSENBERG 1991).

Most recently, clinical research projects have involved the ex vivo infection of patient tumor cells with retroviral vectors encoding either IL-2 or TNF (ROSENBERG et al. 1992a, b). These studies were designed to introduce immunomodulatory cytokine genes into the tumor cells themselves to try to compensate for the toxic side effects of intravenous cytokine infusion, as well as to enhance the local immune response to the tumor. Results are pending in these studies.

In another human trial, a small number of patients with metastatic melanomas and subcutaneous tumor nodules were treated with a DNA vaccine encoding the human MHC class I gene, HLA-B7. These studies were employed to enhance the immunogenicity of these tumors by increasing tumor antigen presentation via HLA-B7 molecules. HLA-B7 expression could be detected in biopsies from patients who had either direct intratumor inoculation or catheter injection into the lung of this gene complexed to a cholesterol-based liposome vector (NABEL et al. 1993, 1994a). These injections appeared to cause no harmful side effects, and led to the production of tumor-specific CTLs in the blood of three out of five patients. One patient who received two treatments plus catheter-based gene delivery showed tumor regression of the treated nodule, as well as complete regression of metastatic tumors at distant sites (NABEL et al. 1994b). The results of these human trials provide great hope for the future development of DNA vaccination protocols for the therapeutic treatment of cancer.

6 Conclusions

DNA vaccination offers several advantages for the therapeutic or prophylactic treatment of cancer. The plasmid vector itself is neither antigenic nor does it have the potential for pathogenicity, in contrast to other viral vaccination vectors, such as adenoviruses, retroviruses, and/or poxviruses. Plasmid DNA delivery also offers the potential for prolonged intracellular gene expression. The long-term persistence of plasmid DNA within vaccinated skeletal muscle allows for the continued stimulation of the immune system and the generation of long-lasting immunity (WOLFF et al. 1990; BRIGHT et al. 1996). Endogenous expression of antigen provides the advantage of presentation by MHC class I antigen, which favors the production of anti-tumor CTL responses. Finally, and importantly production of DNA vaccines is a relatively easy and inexpensive approach compared to other techniques for the development of recombinant, live or attenuated vaccines (Fig. 1).

Several studies described in this review have demonstrated the ability of DNA vaccines to induce both humoral and cell-mediated immunity specific for certain tumor-associated antigens. Furthermore, DNA vaccines encoding cytokine genes

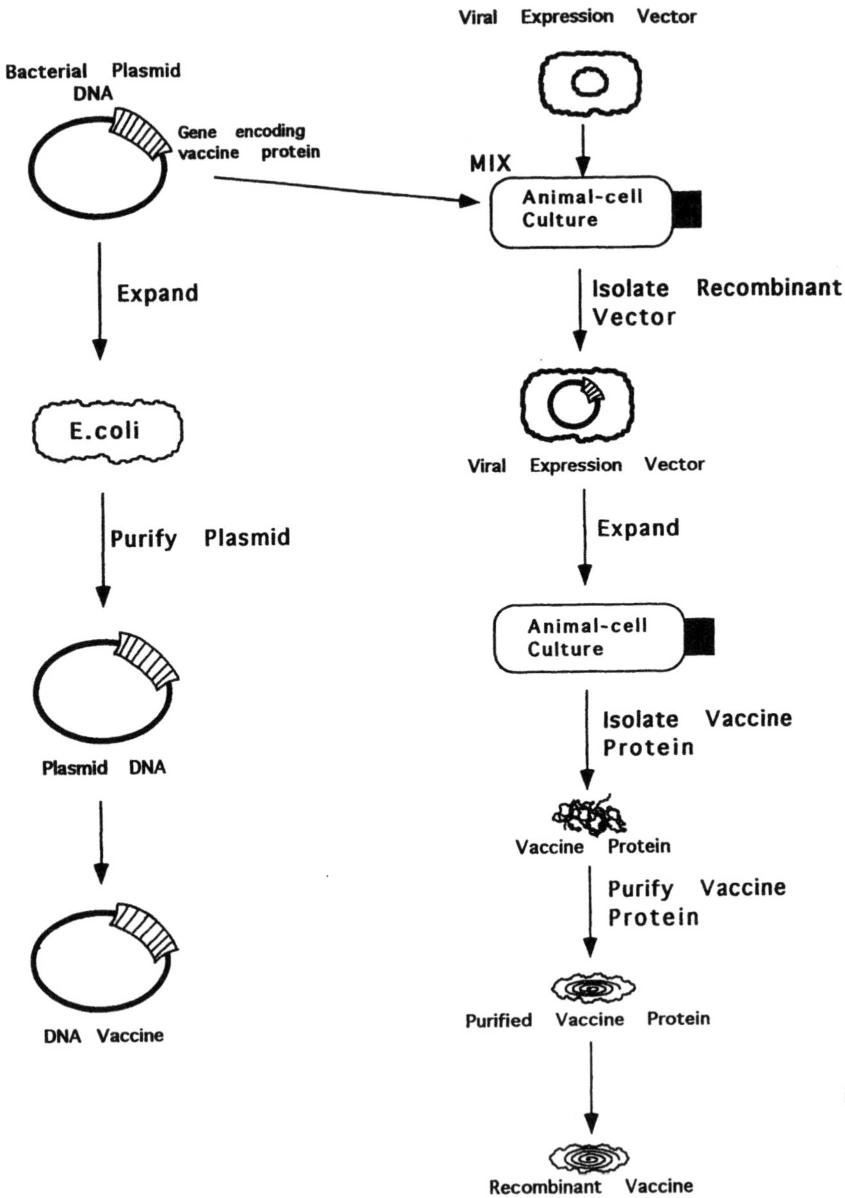

Fig. 1. A comparison of the steps involved in the production of a DNA vaccine (*left panel*) and a recombinant protein vaccine (*right panel*). Initial steps involve the generation of plasmid DNA encoding the protein gene of interest for both types of vaccines. For DNA vaccines, only expansion and purification of the plasmid is required prior to immunization. The production of recombinant protein vaccines requires additional steps for the expression and purification of the protein itself, a much more laborious process when compared to the preparation of a DNA vaccine

or MHC class I genes can stimulate local immune responses through the recruitment of immune cells to the tumor site. Additional research is warranted in this field, because so little is known about the mechanism(s) by which the immune system is stimulated to produce tumor-specific antibodies or CTLs following DNA vaccination. The few clinical trials that have been carried out have not displayed 100% efficiency in tumor regression, so there is more to be learned regarding the effectiveness and mechanism(s) of these approaches. It is clear, however, that the field of genetic vaccination research offers opportunity for the development of safe prophylactic and therapeutic vaccines that are capable of generating tumor-specific protective immunity or tumor regression, respectively.

References

Anderson JL, Martin RG, Chang C, Mora PT and Livingston DM (1977) Nuclear preparations of SV40 transformed cells contain tumor specific transplantation antigen activity. Virology 76:420–425

Bright RK, Beames B, Shearer MH and Kennedy RC (1996) Protection against a lethal tumor challenge with simian virus 40 transformed cells by the direct injection of DNA encoding SV40 large tumor antigen. Cancer Res 56:1126–1130

Bright RK, Shearer MH and Kennedy RC (1994a) Immunization of BALB/c mice with recombinant simian virus 40 large tumor antigen induces antibody-dependent cell-mediated cytotoxicity against simian virus 40-transformed cells. J Immunol 153:2064–2071

Bright RK, Shearer MH and Kennedy RC (1994b) SV40 large tumor antigen associated synthetic peptides define native antigenic determinants and induce protective tumor immunity in mice. Mol Immunol 29:989–999

Browning MJ and Bodmer WF (1992) MHC antigens and cancer: Implications for T-cell surveillance. Curr Opinion Immunol 4:613–618

Carrel S and Rimoldi D (1993) Melanoma associated antigens. Eur J Cancer 29A:1903–1907

Chang C, Martin RG, Livingston DM, Luborsky SW, Hu CP and Mora PT (1979) Relationship between T-antigen and tumor-specific transplantation antigen in simian virus 40 transformed cells. J Virol 29:69–75

Cheever MA, Britzmann-Thompson D, Klarnet JP and Greenberg PI (1986) Antigen-driven long term-cultured T cells proliferate in vivo, distribute widely, mediate specific tumor therapy, and persist long-term as functional memory T cells. J Exp Med 163:1100–1112

Chen L, Ashe S, Brady WA, Hellstrom I, Hellstrom KE, Ledbetter JA, McGowan P and Linsley PS (1992) Costimulation of antitumor immunity by the B7 counterreceptor for the T lymphocyte molecules CD28 and CTLA-4. Cell 71:1093–1102

Ciernik IF, Berzofsky JA and Carbone DP (1996) Induction of cytotoxic T lymphocytes and antitumor immunity with DNA vaccines expressing single T cell epitopes. J Immunol 156:2369–2375

Cohen J (1993) Cancer vaccines get a shot in the arm. Science 262:841–843

Conry RM, LoBuglio AF, Kantor J, Schlom J, Loechel F, Moore SE, Sumerel LA, Barlow DL, S. A and Curiel DT (1994) Immune response to a carcinoembryonic antigen polynucleotide vaccine. Cancer Res 54:1164–1168

Corr M, Lee DJ, Carson DA and Tighe H (1996) Gene vaccination with naked plasmid DNA: mechanism of CTL priming. J Exp Med 184:1555–1560

Cristiano RJ, Smith LC, Brinkley BR and Woo SL (1993a) Hepatic gene therapy: Efficient gene delivery and expression in primary hepatocytes utilizing a conjugated adenovirus-DNA complex. Proc Natl Acad Sci USA 90:11548–11552

Cristiano RJ, Smith LC and Woo SLC (1993b) Hepatic gene therapy: Adenovirus enhancement of receptor-mediated gene delivery and expression in primary hepatocytes. Proc Natl Acad Sci USA 90:2122–2126

Davis HL, Michel ML and Whalen RG (1995) Use of plasmid DNA for direct gene transfer and immunization. Ann New York Acad Sci 772:21–29

Dranoff G and Mulligan RC (1995) Gene transfer as cancer therapy. Adv Immunol 58:417–454

Eisen HN, Sakato N and Hall SJ (1975) Myeloma proteins as tumor-specific antigens. Transplant Proc 7:209–214

Fearon ER, Pardoll DM, Itaya T, Golumbek P, Levitsky HI, Simons JW, Karasuyama H, Vogelstein B and Frost P (1990) Interleukin-2 production by tumor cells bypasses T helper function in the generation of an antitumor response. Cell 60:397–403

Foecking MK and Hofstetter H (1986) Powerful and versatile enhancer-promoter unit for mammalian expression vectors. Gene 45:101–105

Foley EJ (1953) Antigenic properties of methylcholanthrene-induced tumor in mice of the strain of origin. Cancer Res 13:835–837

Gansbacher B, Bannerji R, Daniels B, Zier K, Cronin K and Gilboa E (1990a) Retroviral vector-mediated g-interferon gene transfer into tumor cells generates potent and long lasting antitumor immunity. Cancer Res 50:7820–7825

Gansbacher B, Zier K, Daniels B, Kronin K, Bannerji R and Gilboa E (1990b) Interleukin-2 gene transfer into tumor cells abrogates tumorigenicity and induces protective immunity. J Exp Med 172:1217–1224

Gooding LR (1977) Specificities of killing by cytotoxic lymphocytes generated in vivo and in vitro to syngeneic SV40 transformed cells. J Immunol 118:920–927

Gross L (1943) Intradermal immunization of C3H mice against a sarcoma that originated in an animal of the same line. Cancer Res 3:326–333

Hassett DE and Whitton JL (1996) DNA immunization. Trends in Microbiology 4:307–312

Hawkins RE, Winter G, Hamblin TJ, Stevenson FK and Russell SJ (1993) A genetic approach to idiotypic vaccination. J Immunother 14:273–278

Hawkins RE, Zhu D, Ovecka M, Winter G, Hamblin TJ, Long A and Stevenson FK (1994) Idiotypic vaccination against human B-cell lymphoma. Rescue of variable region gene sequences from biopsy material for assembly as single-chain antibodies Fv personal vaccines. Blood 83:3279–3288

Hui K, Grosveld F and Festenstein (1984) Rejection of transplantable leukemia cells following MHC DNA-mediated cell transformation. Nature 311:750–752

Irvine KR, Rao JB, Rosenberg SA and Restifo NP (1996) Cytokine enhancement of DNA immunization leads to effective treatment of established pulmonary metastases. J Immunol 156:238–245

Kawakami Y, Robbins PF, Wang RF, Rosenberg SA (1996) Identification of tumor-regression antigens in melanoma. In: DeVita VT, Hellman S, Rosenberg SA (eds) Advances in oncology, Lippincott-Raven, Philadelphia, pp 3–21

Kennedy RC, Dreesman GR, Butel JS and Lanford RE (1985) Suppression of in vivo tumor formation induced by simian virus 40-transformed cells in mice receiving anti-idiotypic antibodies. J Exp Med 161:1432–1449

Kennedy RC, Zhou E-M, Lanford RE, Chanh TC and Bona CA (1987) Possible role of anti-idiotypic antibodies in the induction of tumor immunity. J Clin Invest 80:1217–1224

Knowles BB, Koncar M, Pfizenmaier, K., Solter D, Aden DP and Trinchieri G (1979) Genetic control of the cytotoxic T cell response to SV40 tumor associated specific antigen. J Immunol 122:1798–1806

Krieg P, Amtmann E, Jonas D, Fisher H, Zang K and Sauer G (1981) Episomal simian virus 40 genomes in human brain tumors. Proc Natl Acad Sci USA 10

Mernaugh RL, Shearer MH, Bright RK, Lanford RE and Kennedy RC (1992) Idiotypic network components are involved in the murine immune response to simian virus 40 large tumor antigen. Cancer Immunol Immunother. 35:113–118

Moller P and Hammerling G (1992) The role of surface HLA-A, B, C molecules in tumor immunity. Cancer Surg 13:101–128

Mule JJ, Shu S, Schwarz SL and Rosenberg SA (1984) Adoptive immunotherapy of established pulmonary metastases with LAK cell and recombinant interleukin-2. Science 225:1487–89

Nabel EG, Yang Z-Y, Muller D, Chang AE, Gao X, Huang L, Cho KJ and Nabel GJ (1994a) Safety and toxicity of catheter gene delivery to the pulmonary vasculature in a patient with metastatic melanoma. Hum Gene Ther 5:1089–1094

Nabel GJ, Nabel EG, Yang Z, Fox BA, Plautz GE, Gao X, Huang L, Shu S, Gordon D and Chang AE (1994b) Molecular genetic interventions for cancer. Cold Spring Harbor Symposia on Quantitative Biology 59:699–707

Nabel GJ, Nabel EG, Yang Z-Y, Fox BA, Plautz GE, Gao X, Huang L, Shu S and Chang AE (1993) Direct gene transfer with DNA-liposome complexes in melanoma: expression, biologic activity, and lack of toxicity in humans. Proc Natl Acad Sci USA 90:11307–11311

Pass HI, Kennedy RC and Carbone M (1996) Evidence for and implications of SV40-like sequences in human mesotheliomas. In: Important advances in oncology 1996. Devita VT, Hellman S, Rosenberg SA, eds. Lippincott-Raven, Philadelphia pp 89–108

Plautz GE, Yang ZY, Gao X, Huang L and Nabel GJ (1993) Immunotherapy of malignancy by in vivo gene transfer into tumors. Proc Natl Acad Sci USA 90:4645

Prehn RT and Main JM (1957) Immunity to methylcholanthrene-induced sarcomas. J Natl Cancer Inst 18:769–778

Raz E, Carson DA, Parker SE, Parr TB, Abai AM, Aichinger G, Gromkowski SH, Singh M, Lew D, Yankauckas MA, Baird SM and Rhodes GH (1994) Intradermal gene immunization: the possible role of DNA uptake in the induction of cellular immunity to viruses. Proc Natl Acad Sci USA 91:9519–9523

Rosenberg SA (1991) Immunotherapy and gene therapy of cancer. Cancer Res (Suppl) 51:5074s–5079s

Rosenberg SA, Anderson WF, Asher AL, Blaese MR, Ettinghausen SE, Hwu P, Kasid A, Mule JJ, Parkinson DR, Schwartzentruber DJ, Topalian SL, Weber JS, Yannelli JR, Yang JC and Linehan WM (1992a) Immunization of cancer patients using autologous cancer cells modified by insertion of the gene for interleukin-2. Human Gene Therapy 3:75–90

Rosenberg SA, Anderson WF, Asher AL, Blaese MR, Ettinghausen SE, Hwu P, Kasid A, Mule JJ, Parkinson DR, Schwartzentruber DJ, Topalian SL, Weber JS, Yannelli JR, Yang JC and Linehan WM (1992b) Immunization of cancer patients using autologous cancer cells modified by insertion of the gene for tumor necrosis factor. Human Gene Therapy 3:57–73

Rosenberg SA, Lotze MA, Muul LM, Leitman S, Chang AE, Ettinghausen SE, Matory YL, Skibber JM, Shiloni E, Vetto JT, Seipp CA, Simpson C and Reichert CM (1985) Observations on the systemic administration of autologous lymphokine-activated killer cells and recombinant interleukin-2 to patients wiht metastatic cancer. N Eng J Med 313:1485–1492

Rosenberg SA, Lotze MT, Yang JC, Aebersold PA, Linehan WM, Seipp CA and White DE (1989) Experience with the use of high-dose interleukin-2 in the treatment of 652 patients with cancer. Ann Surg 210:474–485

Rosenberg SA, Lotze MT, Yang JC, Topalian SL, Chang AE, Schwartzentruber DJ, Aebersold P, Leitman S, Linehan WM and Seipp CAea (1993) Prospective randomized trial of high-dose interleukin-2 alone or in conjunction with lymphokine-activated killer cells for the treatment of patients with advanced cancer. J Natl Cancer Inst 85:622–632

Rosenberg SA, Packard BS, Aebersold PM, Solomon D, Topalian SL, Toy ST, Simon P, Lotze MT, Yang JC, Seipp CA, Simpson C, Carter C, Bock S, Schwartzentruber D, Wei JP and White DE (1988) Use of tumor-infiltrating lymphocytes and interleukin-2 in the immunotherapy of patients with metastatic melanoma. N Engl J Med 319:1676–1680

Rosenberg SA, Spiess P and Lafreniere R (1986) A new approach to the adoptive immunotherapy of cancer with tumor-infiltrating lymphocytes. Science 233:1318–1321

Roth C, Rochlitz C and Kourilsky P (1994) Immune response against tumors. Adv Immunol 57:281–351

Samson JH, Archer GE, Ashley DM, Fuchs HE, Hale LP, Dranoff G and Bigner DD (1996) Subcutaneous vaccination with irradiated, cytokine-producing tumor cells stimulates $CD8^+$ cell-mediated immunity against tumors located in the "immunologically privileged" central nervous system. Proc Natl Acad Sci USA 93:10399–10404

Schirmbeck R, Bohm W and Reimann J (1996) DNA vaccination primes MHC class I-restricted, simian virus 40 large tumor antigen-specific CTL in $H\text{-}2^d$ mice that reject syngeneic tumors. J Immunol 157:3550–3558

Schirmbeck R, Zerrahn J, Kuhrober A, Kury E, Deppert W and Reimann J (1992) Immunization with soluble simian virus 40 large T antigen induces a specific response of $CD3^+$ C4-$CD8^+$ cytotoxic T lymphocytes in mice. Eur J Immunol 22:759–766

Schwartz RH (1992) Costimulation of T lymphocytes: the role of CD28, CTLA-4, and B7/BB1 in interleukin-2 production and immunotherapy. Cell 71:1065–1068

Shearer MH, Bright RK and Kennedy RC (1993) Comparison of the humoral immune responses and tumor immunity in mice immunized with recombinant SV40 large tumor antigen and a monoclonal anti-idiotype. Cancer Res 53:5734–5739

Stevenson FK, Zhu D, King CA, Ashworth LJ, Kumar S and Hawkins RE (1995) Idiotypic DNA vaccines against B-cell lymphoma. Immunol Rev 145:211–228

Tanaka K, Isselbacher KJ, Khoury G and Jay G (1985) Reversal of oncogenesis by the expression of H-2 K antigen following H-2 gene transfection. Science 228:26–30

Teng MN, Park BH, Koeppen HK, Tracey KJ, Fendly BM and Schreiber H (1991) Long-term inhibition of tumor growth by tumor necrosis factor in the absence of cachexia or T-cell immunity. Proc Natl Acad Sci USA 88:3535–3539

Tepper RI, Pattengale PK and Leder P (1989) Murine interleukin-4 displays potent antitumor activity in vivo. Cell 57:503–512

Tevethia SS, Flyer DC and Tjian R (1980) Biology of simian virus 40 (SV40) transplantation antigen (TrAg). VI. Mechanism of induction of SV40 transplantation immunity in mice by purified SV40 T antigen (D2 protein). Virology 107:13–23

Townsend SE and Allison JP (1993) Tumor rejection after direct costimulation of $CD8^+$ T cells by B7-transfected melanoma cells. Science 259: 368–270

Trail P, Wilner D, Lasch SJ, Henderson AJ, Hofstead S, Casazza AM, Firestone RA, Hellstrom I and Hellstrom KE (1993) Cure of xenografted human carcinomas by BR96-doxorubucin immunoconjugates. Science 261:212–215

Ulmer JB, Donnelly JJ and Liu MA (1996) DNA vaccines promising: a new approach to inducing protective immunity. ASM news 62:476–479

Ulmer JB, Donnelly JJ, Parker SE, Rhodes GH, Felgner PL, Dwarki VJ, Gromkowski SH, Deck RR, DeWitt CM, Friedman A, Hawe LA, Leander KR, Martinez D, Perry HC, Shiever JW, Montgomery DL and Liu MA (1993) Heterologous protection against influenza by injection of DNA encoding a viral protein. Science 259:1745–1749

Vile RG and Hart IR (1993) In vitro and vivo targeting of gene expression to melanoma cells. Cancer Res 53:962–967

Vivetta ES, Thorpe PE and Uhr JW (1993) Immunotoxins: magic bullets or misguided missiles. Immunol Today 14:252–259

Wagner E, Plank C, Zatloukal K, Cotten M and Birnsteiel ML (1992a) Influenza virus hemagglutinin HA-2 N-terminal fusogenic peptides augment gene transfer by transferrin-polylysine-DNA complexes: Toward a synthetic virus-like gene-transfer vehicle. Proc Natl Acad Sci USA 89:7934–7938

Wagner E, Zatloukal K, Cotten M, Kirlappos H, Mechtler K, Curiel DT and Birnsteil ML (1992b) Coupling of adenovirus to transferrin-polylysine/DNA complexes greatly enhances receptor-mediated gene delivery and expression of transfected genes. Proc Natl Acad Sci USA 89:6099–6103

Wallich R, Bulbuc N, Hammerling G, Katzav S, Segal S and Feldman M (1985) Abrogation of metastatic properties of tumor cells by *de novo* expression of H-2 K antigen following H-2 gene transfection. Nature 315:301–305

Wang B, Merva M, Dang K, Ugen KE, Williams WV and Weiner DB (1995) Immunization by direct DNA inoculation induces rejection of tumor cell challenge. Human Gene Ther 6:407–418

Watanabe Y, Kuribayashi K, Miyatake S, Nishihara K, Nakayama E, Taniyama R and Sakata T (1989) Exogenous expression of mouse interferon g cDNA in mouse neuroblastoma C1300 cells results in reduced tumorigenicity by augmented anti-tumor immunity. Proc Natl Acad Sci USA 86:9456–9460

Whalen RG (1996) DNA vaccines for emerging infectious diseases: what if? Emerg Inf Dis 2:168–175

Wolff JA, Malone RW, Williams P, Chong W, Acsadi G, Jani A and Felgner PL (1990) Direct gene transfer into mouse muscle in vivo. Science 247:1465–1468

Xiang Z and Ertl HC (1995) Manipulation of the immune response to a plasmid-encoded viral antigen by coinoculation with plasmids expressing cytokines. Immunity 2:129–135

Zhu N, Liggitt D, Liu Y and Debs R (1993) Systemic gene expression after intravenous DNA delivery into adult mice. Science 261:209–211

Zimmerman W and Thompson J (1990) Recent developments concerning the carcinoembryonic gene family and their clinical applications. Tumor Biol 11:1–4

Vaccination Against Pathogenic Cells by DNA Inoculation

B. Wang[1], A.P. Godillot[2,4], M.P. Madaio[2], D.B. Weiner[3,4], and W.V. Williams[2,4,5,*]

1 Introduction

The goal of vaccination is to induce immunity to protect the host from disease. Vaccines should generate long-term protective immune responses which perform immune surveillance against specific antigens. Currently, a wide spectrum of vaccines are under development against not only infectious diseases, but also against cancers as well as allergic and autoimmune diseases. The mechanisms by which vaccines elicit protective immune responses against tumor growth have not been completely understood. Costimulatory molecule activation and strong cytolytic T cells (CTLs) have been implicated in the control of tumor cell growth or metastasis. Specific monoclonal antibodies have been also shown to control tumor cell growth

[1]Sinogen Institute, Institute of Zoology, Chinese Academy of Sciences, Beijing 100080, Peoples Republic of China

[2]Department of Medicine, Rheumatology Division, University of Pennsylvania School of Medicine, 912 Stellar-Chance Laboratories, 422 Curie Blvd, Philadelphia, PA 19104, USA

[3]The Department of Pathology and Laboratory Medicine, University of Pennsylvania School of Medicine, 505 Stellar-Chance Laboratories, 422 Curie Blvd, Philadelphia, PA 19104, USA

[4]Department of Biotechnology, Institute for Biotechnology and Advanced Molecular Medicine, University of Pennsylvania School of Medicine, Philadelphia, PA 19104, USA

[5]Childrens' Hospital of Philadelphia, Philadelphia, PA 19104, USA

*Corresponding author

to some degree. To achieve protective immune responses against tumor cells, we need to understand the context of the different cellular, humoral and molecular functions of the immune system.

Cellular immunity can be divided into MHC class I-specific and MHC class II-specific. These differ in their activation mechanisms and their effector functions. CD4+ T cells are activated by antigens processed and presented in the MHC class II pathway. The MHC class II pathway is quite efficient at presenting exogenous antigens (those aquired by endocytosis) which are typically extracellular proteins. Exogenous antigens (Ag) are processed in endosomes into antigenic peptides which associate with MHC class II molecules. These antigenic peptides are presented in the context of the MHC class II structure (along with accessory molecules such CD80/CD86) to MHC(II)-restricted, antigen-specific CD4+ T cells. The specific recognition event of the T cell antigen receptor (TCR)/CD3 complex binding to the Ag-MHC(II) complex is essential for T cell activation. CD4+ cells participate in cognate help to activate Ag-specific B cell antibody production and secrete cytokines which vary for distinct CD4+ T cells. These have been subdivided into T_{H0}, T_{H1} and T_{H2} profiles. T_{H1} T cells secrete cytokines including interferon-γ (IFN-γ) and interleukin-2 (IL-2), which activate additional T cells and induce MHC class II expression. T_{H2} cells secrete cytokines such as IL-4 and IL-10, which activate B cells and induce antibody secretion. T_{H0} cells have an intermediate or overlapping phenotype. Thus, the class II pathway induces immunostimulatory responses.

The MHC class I pathway presents endogenously synthesized protein antigens, which have also been processed into peptides. These arise (as in viral infection) from translation of proteins on ribosomes either in the cytosol or in the endoplasmic reticulum (ER). Endogenous antigens are processed into peptides and associate with MHC class I molecules in the ER. The Ag-MHC(I) complex is transported to the cell surface and is recognized by the TCR on CD8+ T cells which frequently have cytotoxic and suppressor activities. Activated CD8+ CTLs preferentially recognize cells which present their cognate Ag-MHC(I) complex and kill these cells. Again, the interaction between the TCR and Ag-MHC determines the specificity of the interaction. CD8+ suppressor cells (Ts) induce states of anergy, or unresponsiveness. Thus, the class I MHC pathway induces cytotoxic and immunosuppressive activities. The different pathways of antigen processing and presentation are depicted in Fig. 1.

In this context, the pathway of antigen presentation is responsible for the characteristics of the immune response. This information can be applied to rational design of vaccines. Such rational approaches should build on prior experience with human vaccination. The first effective vaccine developed was a live/attenuated vaccine, vaccinia. This remains the only vaccine which has successfully eradicated a pathogenic virus from the human population. Interestingly, live/attenuated vaccines engage both the MHC class I and class II pathways. This is due to both endogenous synthesis of virus proteins and processing and presentation of secreted viral proteins. In contrast, both killed vaccine preparations and recombinant protein formulations appear more efficient at inducing MHC class II-restricted responses. This has the consequence that responses are limited to B cell and CD4+

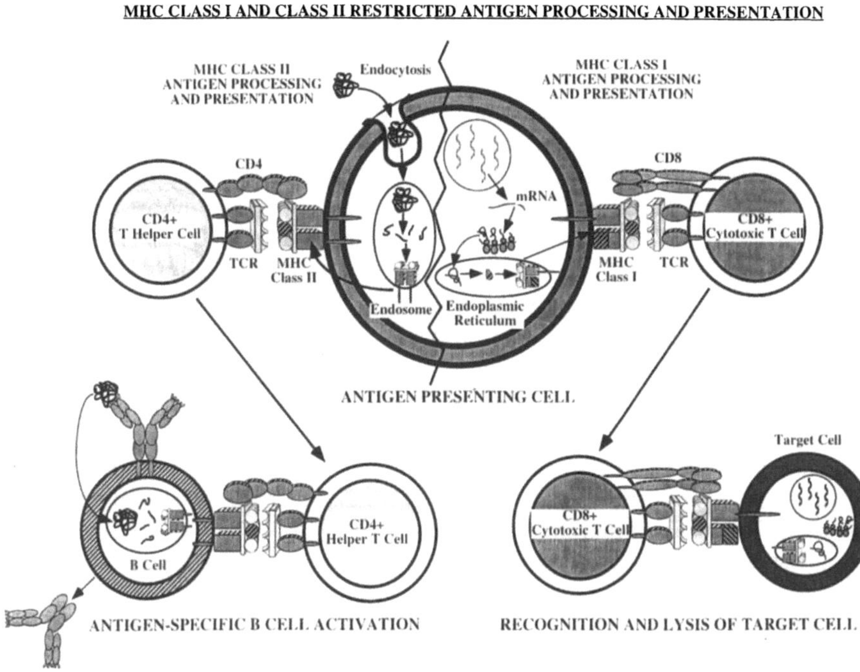

Fig. 1. Antigen processing and presentation pathways. Exogenous antigens (*left side*) are internalized by antigen presenting cells (APC) into endosomes where they are proteolytically degraded into antigenic peptides which associate with MHC class II molecules. The Ag-MHC II complex is presented to CD4+ T cells, where it is recognized by their T cell receptor (TCR). Cross-linking the TCR by Ag-MHC II leads to T cell activation. If T_{H1} cells are activated, they in turn can activate antigen-specific B cells to make antibody. If T_{H2} cells are activated, they stimulate a cellular response. Endogenous antigens (*right side*) are made in the cell and processed into peptides in the cytoplasm. They are transported into the endoplasmic reticulum where they associate with MHC class I molecules. The Ag-MHC I complex is presented to CD8+ T cells, where it is recognized by their T cell receptor (TCR). Cross-linking the TCR by Ag-MHC I leads to T cell activation. The activated CD8+ T cells can be directly cytotoxic to cells expressing the Ag-MHC I complex, or can serve to suppress immune responses

responses, including antibody formation and delayed-type hypersensitivity responses (such as tuberculin skin test responses), but is less efficient for inducing the CD8+ component. What is lacking at present from the current therapeutic vaccine repertoire is a method of preferentially inducing class I-restricted CD8+ T cell responses without the risk of an active infection, such as an attenuated vaccine produces.

One method of inducing class I-restricted responses is to develop a vaccine that utilizes the intracellular processing pathway. This can be developed by introducing specific genes into cells, allowing the endogenous synthesis of their products. Currently, at least three methods have been developed that predominately utilize the class I pathway: recombinant virus infection, "abortive infection" using defective retroviruses and gene inoculation. Recombinant viruses are genetically modified viruses which include the gene(s) encoding the antigen of interest. These recombinant viruses also encode viral antigens, so immune responses to multiple antigens are elicited. In retroviral vaccination, a gene with an appropriate promoter sequence is transcribed with a specific packaging signal that allows packaging of the transcript in a retroviral particle which lacks viral transcripts. Such packaged transcripts are delivered to appropriate target cells by abortive infection in vivo. Following reverse transcription, the cDNA is transcribed and the transcript directs synthesis of the packaged gene product into the endogenous pathway. Similarly, gene inoculation introduces a genetic sequence under control of a specific promoter into cells in vivo. In this system, no packaging signal is needed. By these techniques, entry into the exogenous pathway can be directed by the addition of appropriate leader peptides on the packaged gene product.

Rational utilization of these techniques should allow definition of those components of a specific immune response which are protective for a given clinical situation. In cancer, target tumor antigens may be of great utility in eliciting protective immune responses. Currently, several approaches for tumor immunotherapy are being developed. One relies on the identification of tumor-associated antigens (TAAs) to be incorporated into the vaccine; another involves increasing the immunogenicity of tumor cell-based vaccines. The latter approach has been increasingly popular because it requires neither identification of TAAs nor the production of that antigen for vaccination. This approach includes the administration of lymphokines to induce anti-tumor effects (Lafreniere and Rosenberg 1985a, b), the introduction of cytokine genes into tumor cells to provide local T cell help (Dranoff et al. 1993; Tepper et al. 1989), the transfer of MHC class I genes into tumor cells to induce specific CTLs (Nabel et al. 1993; Plautz et al. 1993) and the induction of tumor differentiation by fusion with normal lymphocytes (Guo et al. 1989). However, this method requires customized therapy for each patient. In contrast, once a TAA is identified that is common to a group of tumors, vaccination against the TAA has many advantages such as specificity and practicality as well as more general administration. A novel technique has been developed that combines two approaches by using a fusion protein made between a TAA and granulocyte/macrophage colony-stimulating factor (GM-CSF); this protein elicits protective immune responses against a tumor challenge in a murine model system (Tao and Levy 1993). The methodology requires production and purification of the fusion proteins, which is costly and time-consuming. Another limitation of protein immunization-based strategies is that they stimulate specific anti-TAA CTLs very weakly. Genetic techniques have been investigated which allow induction of antitumor CTLs and other arms of the immune system.

2 Retroviral Vaccination

Gene therapy techniques frequently require a genetically engineered nonpathogenic viral infection event for gene transduction. In theory such techniques mimic aspects of abortive infection events in that the genetic sequences deliver an exogenous antigen which can then be transcribed and translated by the host cells' endogenous molecular machinery. Such transcription may allow delivery of the vector encoded protein for presentation by both class I and class II MHC antigen systems resulting in specific and relevant cellular and humoral immunity. Studies using a mouse retroviral infection system demonstrated that drug interference with initial oncogenic retroviral infection in vivo results in protection from subsequent oncogene retroviral challenge and development of viremia. Further studies reported the induction of both CD4+ and CD8+ cells during the development of protective immune responses (Hom et al. 1991; Ruprecht et al. 1990). These results indicate that if a retrovirus can be prevented from replicating within the host, the benifits of a live attenuated viral immunization can still be achieved. One alternative, therefore, is to engineer a replication-defective retrovirus or use gene sequences directly for immunization. A further advantage for human vaccine design strategies would be the existence of a reproducible small animal system for evaluation of any potential immune responses through a survival-based challenge.

We reported the successful production of B cell, T helper as well as T cytotoxic immune responses through abortive retroviral infection of live animals (Williams et al. 1993). To evaluate a retroviral transduction immunization strategy in vivo, we utilized a mouse model tumor system. The human and mouse CD4 protein share both structural as well as some primary amino acid similarity. Accordingly the human CD4 gene is a unique model for the generation of immune responses to the immunoglobulin supergene family.

Immunoglobulins and TCRs share structural and primary amino acid sequences yet are clonotypically distinct entities within the immunoglobulin super gene family which includes not only host-encoded but also viral proteins (Kieber-Emmons et al. 1989). We therefore utilized the human CD4 molecule to test the ability of retroviral immunization to induce relevant immune responses to this model antigen. Immune competent Balb/c mice which are injected with SP2/0 B lymphoid tumor cells or human CD4 gene marked lymphoma cells develop lethal tumor-induced disease. In contrast mice retrovirally transduced with a replication defective retroviral vector containing the functional human CD4 cDNA sequence generated a protective immune response, with survival of 50% of the vaccinated mice and prolonged survival in the remaining mice. In contrast, unvaccinated mice or mice vaccinated with a replication defective retroviral vector that expressed an irrelevant antigen always succumbed to tumor challenge within 8 weeks. Analysis of immune responses in vaccinated animals demonstrated the presence of both B cell and T cell responses, as evidenced by a specific humoral immune response, cellular proliferative responses and cytotoxic responses. These studies indicate that retroviral immunization was capable of eliciting both humoral and cellular immunity to human CD4, and this resulted in a protective immune response to SP2/0-CD4 B lymphoma cells. These studies encouraged investigation of the feasibility of using additional, simpler and more easily applicable strategies for the development of similar responses.

3 DNA Inoculation

Direct immunization by DNA inoculation (gene immunization) appears to mimic some aspects of attenuated vaccines in that synthesis of specific foreign proteins would be accomplished in the host and be the subject of immune surveillance. Gene immunization is dependent upon injection of a nucleic acid sequence directly into a host target tissue, such as muscle (NABEL et al. 1990; WOLFF et al. 1990). Initially investigations demonstrated gene expression as the result of intramuscular injection of the human growth hormone (hGH) gene as well as the muscular dystrophy gene in mice (NABEL et al. 1990; WOLFF et al. 1990). This can be antigenic in some circumstances and can elicit an antibody-based immune response (TANG et al. 1992). Using a variety of delivery methods, DNA constructs directly inoculated into animals have been shown to generate host immunity against influenza virus, HIV-1, bovine hepatitis virus, human hepatitis B surface antigen, HTLV-I, and human immunoglobulin V regions (AGADJANYAN et al. 1994; CONRY et al. 1994; DAVIS et al. 1993; RAZ et al. 1993; ROBINSON et al. 1993; ULMER et al. 1993; WANG et al. 1993a, b). It is possible that direct inoculation of a TAA gene could induce an anti-tumor cell response that could inhibit tumor growth and lead to tumor rejection.

We were among the first groups to develop the DNA immunization technology to generate humoral and cellular immune responses to the envelope glycoprotein the HIV-1 virus (WANG et al. 1993a, b). In this study, a plasmid construct encoding the entire envelope glycoprotein of HIV-1 was injected intramuscularly into mice. These mice produced humoral immune responses in which the antibodies were: (a) able to bind to several important epitopes of the envelope glycoprotein and (b) neutralize infection of target cells by cell-free virus. In addition these mice generated CTLs against HIV-1. Interestingly, the humoral anti-envelope peptide immune responses generated after immunization were more extensive than those generated after immunization with the analogous recombinant protein. We have also reported that direct DNA immunization can induce protection from malignant tumor cell challenge through the generation of specific immune responses directed against antigens displayed on the tumor cells (WANG et al. 1995). The protected mice remain tumor-free for more than 1 year post-challenge. Memory responses upon tumor rechallenge were observed for both humoral and cellular immunity. Inoculated animals were able to reject otherwise lethal tumors several months following the original DNA inoculation protocol.

4 DNA Immunization Against Model Tumors

One readily defined TAA is the variable (V) region of the antibody or T cell antigen receptor expressed on lymphomas and leukemias. Several studies have targeted antibody V regions in the induction of anti-tumor immunity (Campbell et al. 1990; Syrengelas et al. 1996; Tao and Levy 1993). We investigated the potential applicability of gene immunization in a pilot system using a model tumor antigen–human CD4. Human CD4 is closely related in sequence to mouse CD4, with 68.458% similarity and 55.374% identity between the two molecules. This degree of similarity is comparable to the similarity between different antibody V regions, and the variability is localized to reverse turn regions similar in structure to antibody CDRs. Human CD4 was stably transfected into SP2/0 cells to develop a tumor challenge system to evaluate the efficacy of gene immunization in vivo. These SP2/0-CD4 cells were passaged through mice to establish CD4+ cell lines that reproducibly induced tumors when inoculated into naive, unmanipulated animals.

A CD4 construct was used to evaluate gene inoculation against these tagged tumor cells. This utilized the pMV7 vector, which employs retroviral long terminal repeats (LTRs) as promoters for CD4 expression. This plasmid was grown, purified, and inoculated into naive Balb/c mice. Following inoculation of 100 μg intramuscularly, the mice were boosted 2 weeks later, and bled the following week. Antibody responses were monitored by ELISA (Wang et al. 1993b). Significant anti-human CD4 antibody levels were detected in these assays. Immunization with this construct, which encodes membrane-associated CD4 molecules, elicited significant antibody responses following two inoculations (Wang et al. 1995). This indicates the ability of antibody-like molecules to elicit antibody responses following gene inoculation. CTL responses were evaluated using a murine fibroblast cell line expressing human CD4. Following immunization, the mice were killed and spleen cells restimulated in vitro with CD4-expressing SP2/0 cells. Following restimulation, the effector cells were washed and evaluated for specific lysis of ^{51}Cr-labeled SP2/0-CD4 cells. The gene immunized animals developed significant cytotoxic T cell activity against the SP2/0-CD4 cells. Thus, using this protocol gene immunization induces effector cells which specifically lyse cells expressing the target antigen.

The ability of gene immunization to protect naive mice from lethal challenge with the marked tumor cells was also investigated. In these studies, groups of four mice were inoculated with 100 μg T4-pMV7 or control plasmid (pMV7) and challenged 1, 2, or 3 weeks later with 3×10^6 SP2/0-CD4 cells. Control mice all developed tumors within 7 weeks and died within 8 weeks. In contrast, no more than 50% of the CD4 gene immunized mice developed tumors or died following 4 months of observation (Wang et al. 1995). In a related experiment, 14 mice were inoculated with T4-pMV7 or pMV7 every week for a total of three inoculations. They were rested for 2 weeks and challenged either with SP2/0-CD4 cells or with SP2/0 cells. All of the mice inoculated with pMV7 developed tumors and died within 8 weeks. Similarly, the mice inoculated with T4-pMV7 and challenged with unmarked SP2/0 cells also succumbed to tumors within 8 weeks. In contrast, 10/10 mice inoculated with T4-pMV7 and challenged with SP2/0-CD4 cells developed no signs of tumors and remained alive for at least 4 months following (Wang et al. 1995). The immunological assays were also repeated on gene immunized mice following challenge with the SP2/0-CD4 cells. There were modest increases in antibody titer to CD4 in mice surviving challenge with SP2/0-CD4 cells, indicating secondary immune responses. The most dramatic alterations were seen in CTL activity, in which specific lysis was evident at several fold lower effector:target cell ratios (WANG et al. 1995). The results indicate a strong class I-mediated response to cellular challenge in this system, suggesting an important role for the MHC class I CD8/CTL system in the development of the observed protective immune response.

Several other investigators have demonstrated the utility of DNA vaccination in developing anti-tumor responses in model systems. CONRY et al. (1994) developed a polynucleotide vaccine against human carcinoembryonic antigen using the cytomegalovirus (CMV) promoter. They used this vaccine to demonstrate that immunization of mice induced humoral and cellular immune responses. Their study also demonstrated the induction of CEA-specific memory T cells by DNA vaccination. A later study by this group demonstrated that CEA DNA vaccination was capable of inducing protection from in vivo tumor challenge (CONRY et al. 1995). They noted that DNA vaccination was dose- and schedule-dependent, with immunizations of 50 μg plasmid DNA inducing better responses than 10 μg, and more frequent immunizations (weekly or thrice weekly inoculations) more effective than less frequent (every third week) immunizations. In their protection studies, they demonstrated that polynucleotide DNA vaccination was capable of eliciting protective responses against syngeneic CEA-transduced colon carcinoma cells as early as 3 weeks after the first vaccination.

In a similar series of experiments, IRVINE et al. (1996) investigated the utility of DNA vaccination against a model tumor antigen using a pulmonary metastatic tumor which expressed the β-galactosidase (β-gal) gene. They demonstrated that epidermal DNA immunization against β-gal induced specific antibody and cytolytic responses. They further demonstrated that DNA immunization against β-gal protected from pulmonary metastatic tumor, and that adoptive transfer of spleenocytes which were activated in vitro reduced the number of established pulmonary nodules. These authors found that while DNA immunization alone has no effect on established pulmonary metastases, significant reduction in the number of established metastases was observed if rIL-2, rIL-6, rIL-7, or particularly rIL-12 were given after DNA inoculation. This indicates the potential for DNA immunization to treat established tumors as well as to prevent tumor formation.

DNA immunization has also been shown to be effective in inducing CTL responses against a single epitope from a mutant form of the human p53 gene (CIERNIK et al. 1996). Using a minigene coding for a single epitope derived from mutant p53, the authors demonstrated that particle bombardment-mediated DNA transfer into the skin of mice induced CTL responses. This CTL response was enhanced if the DNA vaccine was fused in frame with the adenovirus E3 leader sequence, targeting the vaccine to the ER. This indicates that intracellular targeting can have a significant impact on the character of the immune responses elicited by DNA vaccines.

These studies establish the potential utility of DNA vaccination against model tumor antigens. However, all of the antigens used were in fact foreign proteins which typically are much more immunogenic than tumor self-antigens which would be encountered clinically. A recent study by SYRENGELAS et al. (1996) used the V_H and V_L regions of a murine lymphoma in an expression vector also encoding the human $C_{\gamma}1$ and C_{κ} with or without linked expression of human GM-CSF. Thus, their immunogen also had foreign antigenic determinants linked to the self-V regions of interest. They demonstrated that intramuscular or intradermal DNA inoculation with these constructs resulted in an anti-idiotypic antibody response, as well as partial protection from tumor challenge in vivo. The linked expression of human GM-CSF markedly enhanced the responses elicited. They concluded that DNA immunization could induce immune responses against a weak, otherwise unrecognized tumor antigen, but that this was dependent on additional stimuli with the DNA (i.e., the human constant regions and GM-CSF). This additional stimulus represents a form of adjuvant effect (boosting the immune response). The potential use of such "molecular adjuvants" in tumor immunity is stressed here.

5 Molecular Adjuvants

The efficacy of many vaccines can be boosted by the co-administration of adjuvants. While many adjuvants are available experimentally, few have received approval for clinical use. This reflects the poorly defined mechanism of action of these adjuvants. Recently, molecular adjuvants with more clearly defined mechanisms of action have been described. These include cytokines (such as GM-CSF and IL-12) and accessory molecules (e.g., CD80 and CD86). The data available on these adjuvants is summarized here.

5.1 Granulocyte/Macrophage Colony-Stimulating Factor

Recent studies of tumor specific immunity also indicate a critical role for GM-CSF in inducing protective responses. TAO and LEVY (1993) investigated an idiotype/GM-CSF fusion protein as a vaccine for B cell lymphoma. By fusing a tumor-derived idiotype to GM-CSF, it was converted into a strong immunogen capable of inducing idiotype-specific antibodies without other carrier proteins or adjuvants. The Id/GM-CSF fusion protein immunization was capable of protecting recipient animals from challenge with an otherwise lethal dose of tumor cells. This indicates the potency of GM-CSF as an immune stimulator allowing development of tumor-specific immunity. DRANOFF et al. (1993) used a genetic approach to compare the ability of different cytokines and other molecules to enhance the immunogenicity of tumor cells. They generated 10 retroviruses encoding potential immunomodulators and studied the vaccination properties of murine B16 melanoma cells transduced by the viruses. Irradiated tumor cells expressing murine GM-CSF stimulated potent, long-lasting, and specific anti-tumor immunity (DRANOFF et al. 1993). Thus, GM-

CSF appears to be a more potent stimulator of anti-tumor immunity than many cytokines currently under investigation for this purpose. In addition, co-inoculation of the GM-CSF gene with a rabies virus nucleic acid vaccine has been demonstrated to boost immune responses specifically against rabies virus, including helper T cell and antibody responses (Xiang et al. 1995). As discussed above, Syrengelas et al. (1996) have demonstrated that GM-CSF boosts protective immunity of an idiotypic DNA vaccine against lymphoma. Thus, there is ample experimental evidence for the use of GM-CSF gene as a molecular adjuvant for vaccines and in inducing protective in vivo responses.

5.2 Interleukin-12

IL-12 has been the object of intense study in recent years, and these studies indicate that IL-12 enhances several key immune responses, including T_{H1} responses and anti-tumor immunity (Brunda et al. 1996; Trinchieri 1995). The anti-tumor effects of IL-12 have been demonstrated in a number of murine tumor models. These effects include inhibition of established experimental pulmonary or hepatic metastases and a reduction in spontaneous metastases. In some studies, mice cured of their tumors by IL-12 are resistant to rechallenge with the tumor. Both T-cells and IFN-γ induction appear necessary for the antitumor effects of IL-12, but the explicit immune mechanisms have not been defined. (These studies are summarized in Brunda et al. 1996).

We have found that IL-12 boosts immunity to DNA vaccines, particularly CTL responses. We tested a stimulation of CTL induction using IL-12 genes as immune modulators along with HIV gene vaccines. These studies (Kim et al. 1997) indicate that the co-administration of IL-12 gene cassettes along with DNA vaccine formulation for HIV-1 resulted in a significant change in the spleen phenotype. Spleens collected from all immunized animals were weighed and visually examined. Whereas the spleens from the mice injected with single formulation DNA vaccines weighed similar to those of the negative control mice (about 100 mg), the spleens from mice injected with DNA vaccine and IL-12 weighed almost three times as much as the control spleens. It is interesting to note that DNA vaccine only or IL-12 only did not result in significantly enlarged spleens. Only when the antigen and IL-12 gene cassettes were co-injected did the splenomegaly result. In addition, the number of leukocytes from the antigen + IL-12 spleens was more than twice the number from the control spleens. Thus, the co-injection of DNA vaccine and IL-12 genes resulted in the increased size of spleen and corresponding augmentation of the number of leukocytes in the spleen.

IL-12 also markedly enhanced CTL activity induced by DNA vaccination. We used a direct CTL assay wherein CTLs are measured directly from spleenocytes without in vitro restimulation. Inoculation with the IL-12 gene cassette alone resulted in no specific lysis of target cells above the background level. A low level (3%) of specific lysis was observed following DNA vaccine alone at the 50:1 effector:target ratio. In contrast, 62% specific lysis was seen following DNA vaccine + IL-12 co-administration at the 50:1 effector:target ratio, and this titered out to 9% at the 12.5:1 effector:target ratio. These studies indicate that IL-12 can markedly enhance CTL induction by DNA vaccines.

5.3 CD80 and CD86

Several studies have investigated the use of the accessory molecules CD80 (B7-1) and CD86 (B7-2) as molecular adjuvants. CD80 transfection of nonimmunogenic tumor cells renders them immunogenic for the elicitation of CTLs (Gajewski 1996; Liu et al. 1996). Blocking CD80 vs CD86 has been shown to alter the immune phenotype in a reciprocal fashion in an autoimmune model, with blockade of CD80 resulting in a T_{H2} response, while blockade of CD86 results in a T_{H1} response (Kuchroo et al. 1995). Several studies indicate that transfection of CD80 into tumor cells renders them more immunogenic, with resultant protective immunity against the tumor (Coughlin et al. 1995; Gajewski et al. 1996; Townsend et al. 1994). These studies, in combination with the IL-12 studies noted above, suggest that T_{H1} and CTL responses are the most efficacious in inducing anti-tumor immunity. They also suggest potential efficacy of CD80 for DNA co-inoculation with tumor vaccines.

5.4 Future Prospects

Tumor metastasis is a complex and poorly understood process. Novel methods of intervention at various steps of the metastatic cascade are critical since the high mortality associated with malignancy is often not due to the primary lesion but to metastatic disease. The murine B16 melanoma system has been used extensively to study the process of metastasis. It has provided a system which utilizes the colonization of the lungs of syngeneic mice after intravenous tail vein injection of cells as the endpoint assay for metastatic potential. This system has been particularly useful for the analysis of the ability of coadministered agents to inhibit lung colonization (Fidler 1978).

Melanoma cells have been known to express unique tumor antigens that are recognized by the host immune system. One melanoma antigen with a species cross-reactive epitope defined in the B16 melanoma system (syngeneic to the C57Bl/6 mouse) has been determined to be a GM3 ganglioside (Wakabayashi et al. 1981). Several other endogenous self-molecules have been identified as tumor-associated cell surface antigens (Ullrich et al. 1986). Ulrich et al. (1986) have detected a heat shock protein (Hsp84) as a tumor-specific antigen with immunodominant epitopes on a methylcholanthrene-induced fibrosarcoma. Recently, Wakabayashi et al. (1981) have developed monoclonal antibodies (M562 and M662) which react solely with B16 melanoma cells but not normal tissue. Therefore, it appears that even self components in normal cells do work as tumor antigens if they are constitutively expressed or modified on malignant cells.

Hayashi et al. (1992) have recently molecularly characterized, using the above mentioned monoclonal antibodies, a gp80 melanoma antigen on the B16 murine melanoma cell line and identified it as the envelope protein of the endogenous ecotropic murine leukemia virus (EEMuLV). The antigen was identified on a number of murine tumor cell lines, but not on fetal tissue, suggesting a specific association with the transformed phenotype. Furthermore, it was determined that the gene for EEMuLV is defective in normal tissue of the C57Bl/6 mouse. This defect is thought to be responsible for the lack of expression of this protein in normal tissues. Further analysis indicates that the gp80 antigen may possess the immunodominant epitopes important in anti-melanoma cytotoxic T cell responses.

We have obtained a cDNA clone (H52) encoding the EEMuLV env protein of the B16 melanoma murine melanoma cell line from Dr. Hayashi. We have subcloned H52 into one of our DNA vaccine vectors, designated as pcH52. We hypothesize that immunization with the pcH52 can elicit significant humoral and cellular immune responses which will prevent B16 tumor growth in vivo. Codelivery of the molecular adjuvants (such as genes encoding for cytokines) may synergize with the DNA vaccines, eliciting protective immunity. In light of the recent evidence suggesting the efficacy of anti-melanoma vaccines (Morton et al. 1992) in humans, these studies demonstrate the potential anti-cancer and anti-metastatic efficacy of the DNA immunization technique.

These studies indicate the feasibility of developing DNA vaccines targeting tumor cells or pathogenic T or B cell variable regions. However, many questions need to be addressed to further develop this technology. The first is the in vitro correlates of protective immunity. Based on the marked enhancement of CTL activity seen in survivors of CD4-expressing tumor cell challenge, and the implication of CTL responses as critical by several other authors, it is suggested that CTL responses are involved in the protective response. However, this does not rule out a protective role for T_H cells and antibodies in vivo. Secondly, as cytokine profiles have not been extensively evaluated, it is uncertain if T_{H1} vs T_{H2} responses are elicited by DNA vaccines, and how the nature of the antigen impacts on the character of the response elicited. This is an important consideration, as T_{H2} responses may be quite detrimental in developing a vaccine against, for example, an autoreactive B cell. In this situation, T_{H2} responses would increase autoantibody production. These two points indicate the essential need to develop vaccines which elicit specific arms of the immune system and not others, and to test them in a reliable system to see if protective responses are indeed elicited.

These studies also indicate a need to increase the efficacy of the DNA vaccines. Note that the CD4 DNA vaccine elicited protective responses only if given before tumor challenge. In human cancer and autoimmune disease, this is not a viable option as the pathogenic clones are already present at the time the patient is encountered. This indicates a need to increase the potency of the vaccines. Studies of vaccination against genetically engineered tumor cells have revealed enhancement of immunogenicity by modifying the cells to express cytokines and costimulatory molecules such as GM-CSF, IL-2, and CD86 (Dranoff et al. 1993; Wakimoto et al. 1996). These studies suggest that the potency of DNA vaccines may be similarly enhanced. Data from the β-gal system support a role for cytokines in enhancing immunity, and in conjunction with DNA vaccines producing effective immunotherapy against established tumors (Irvine et al. 1996). However, the foreign nature of β-gal as opposed to a tumor antigen (which is functionally a self-antigen) may influence the success of this strategy in the clinic.

Thus, additional studies are needed to increase the potency of DNA vaccines for use in human cancer and autoimmune disease. While these are underway, we have begun a trial of a DNA vaccine in human cutaneous T cell lymphoma, with four patients inoculated to date. These studies will be critical in establishing the safety of DNA vaccines and their potential efficacy prior to embarking on more widespread clinical trials.

References

Agadjanyan MG, Wang B, Ugen KE, Villafana T, Merva M, Petrushina I, Williams WV and Weiner DB (1994) DNA inoculation with an HTLV-I envelope construct elicits immune responses in rabbits. Vaccines 94:47–53

Brunda MJ, Luistro L, Rumennik L, Wright RB, Dvorozniak M, Aglione A, Wigginton JM, Wiltrout RH, Hendrzak JA and Palleroni AV (1996) Antitumor activity of interleukin 12 in preclinical models. Cancer Chemother Pharmacol 38 Suppl:S16–21

Campbell MJ, Esserman L, Byars NE, Allison AC and Levy R (1990) Idiotype vaccination against murine B cell lymphoma. Humoral and cellular requirements for the full expression of antitumor immunity. J Immunol 145:1029–36

Ciernik IF, Berzofsky JA and Carbone DP (1996) Induction of cytotoxic T lymphocytes and antitumor immunity with DNA vaccines expressing single T cell epitopes. J Immunol 156:2369–75

Conry RM, LoBuglio AF, Kantor J, Schlom J, Loechel F, Moore SE, Sumerel LA, Barlow DL, Abrams S and Curiel DT (1994) Immune response to a carcinoembryonic antigen polynucleotide vaccine. Cancer Res 54:1164–8

Conry RM, LoBuglio AF, Loechel F, Moore SE, Sumerel LA, Barlow DL and Curiel DT (1995) A carcinoembryonic antigen polynucleotide vaccine has in vivo antitumor activity. Gene Ther 2:59–65

Coughlin CM, Wysocka M, Kurzawa HL, Lee WM, Trinchieri G and Eck SL (1995) B7-1 and interleukin 12 synergistically induce effective antitumor immunity. Cancer Res 55:4980–7

Davis HL, Michel M-L and Whalen RG (1993) DNA-based immunization induces continuous secretion of hepatitis B surface antigen and high levels of circulating antibody. Human Mol Genetics 2:1847–1851

Dranoff G, Jaffee E, Lazenby A, Golumbek P, Levitsky H, Brose K, Jackson V, Hamada H, Pardoll D and Mulligan RC (1993) Vaccination with irradiated tumor cells engineered to secrete murine granulocyte-macrophage colony-stimulating factor stimulates potent, specific, and long lasting anti-tumor immunity (Meeting abstract). Gene Therapy for Neoplastic Diseases June 26–29, 1993, Washington, DC, A

Fidler IJ (1978) Tumor heterogeneity and the biology of cancer invasion and metastasis. Cancer Res 38:2651–60

Gajewski TF (1996) B7-1 but not B7-2 efficiently costimulates CD8+ T lymphocytes in the P815 tumor system in vitro. J Immunol 156:465–72

Gajewski TF, Fallarino F, Uyttenhove C and Boon T (1996) Tumor rejection requires a CTLA4 ligand provided by the host or expressed on the tumor:superiority of B7-1 over B7-2 for active tumor immunization. J Immunol 156:2909–17

Guo HG, Veronese F, Tschachler E, Pal R, Gallo RC and Reitz MS (1989) Characterization of an HIV-1 point mutation blocked in envelope glycoprotein cleavage. V. International conference on AIDS

Hayashi H, Matsubara H, Yokota T, Kuwabara I, Kanno M, Koseki H, Isono K, Asano T and Taniguchi M (1992) Molecular cloning and characterization of the gene encoding mouse melanoma antigen by cDNA library transfection. J Immunol 149:1223–9

Hom RC, Finberg RW, Mullaney S and Ruprecht RM (1991) Protective cellular retroviral immunity requires both CD4 and CD8 immune T cells. J Virol 65:220–224

Irvine KR, Rao JB, Rosenberg SA and Restifo NP (1996) Cytokine enhancement of DNA immunization leads to effective treatment of established pulmonary metastases. J Immunol 156:238–45

Kieber-Emmons T, Jameson B and Morrow W (1989) The gp120-CD4 interface:structural, immunological and pathological considerations. Biochim Biophys Acta 989:281–300

Kim JJ, Ayyavoo V, Bagarazzi ML, Chattergoon MA, Dang K, Wang B, Boyer JD and Weiner DB (1997) In vivo engineering of a cellular immune response by coadministration of IL-12 expression vector with a DNA immunogen. J Immunol 158:816–826

Kuchroo VK, Das MP, Brown JA, Ranger AM, Zamvil SS, Sobel RA, Weiner HL, Nabavi N and Glimcher LH (1995) B7-1 and B7-2 costimulatory molecules activate differentially the Th1/Th2 developmental pathways:application to autoimmune disease therapy. Cell 80:707–18

Lafreniere R and Rosenberg SA (1985a) Adoptive immunotherapy of murine hepatic metastases with lymphokine activated killer (LAK) cells and recombinant interleukin 2 (RIL 2) can mediate the regression of both immunogenic and nonimmunogenic sarcomas and an adenocarcinoma. J Immunol 135:4273–80

Lafreniere R and Rosenberg SA (1985b) Successful immunotherapy of murine experimental hepatic metastases with lymphokine-activated killer cells and recombinant interleukin 2. Cancer Res 45:3735–41

Liu B, Podack ER, Allison JP and Malek TR (1996) Generation of primary tumor-specific CTL in vitro to immunogenic and poorly immunogenic mouse tumors. J Immunol 156:1117–25

Morton DL, Foshag LJ, Hoon DS, Nizze JA, Famatiga E, Wanek LA, Chang C, Davtyan DG, Gupta RK, Elashoff R and Irie RF (1992) Prolongation of survival in metastatic melanoma after active specific immunotherapy with a new polyvalent melanoma vaccine [published erratum appears in Ann Surg 1993 Mar; 217(3): 309]. Ann Surg 216:463–82

Nabel EG, Plautz G and Nabel GJ (1990) Site-specific gene expression in vivo by direct gene transfer into the arterial wall. Science 249:1285–1288

Nabel GJ, Nabel EG, Yang ZY, Fox BA, Plautz GE, Gao X, Huang L, Shu S, Gordon D and Chang AE (1993) Direct gene transfer with liposome-DNA complexes in melanoma:expression, biological activity and lack of toxicity in humans. Proc Natl Acad Sci USA 90:11307–11311

Plautz GE, Yang ZY, Wu BY, Gao X, Huang L and Nabel GJ (1993) Immunotherapy of malignancy by in vivo gene transfer into tumors. Proc Natl Acad Sci USA 90:4645–4649

Raz E, Watanabe A, Baird SM, Eisenberg RA, Parr TB, Lotz M, Kipps TJ and Carson DA (1993) Systemic immunological effects of cytokine genes injected into skeletal muscle. Proc Natl Acad Sci USA 90:4523–7

Robinson HL, Fynan EF and Webster RG (1993) Use of direct DNA inoculations to elicit protective immune responses. Vaccines 93 Cold Spring Harbor Press:311–315

Ruprecht MR, Mullaney S, Bernard LD, Sosa MAG, Hom RC and Finberg RW (1990) Vaccination with a live retrovirus:The nature of the protective immune response. PNAS 87:5558–5562

Syrengelas AD, Chen TT and Levy R (1996) DNA immunization induces protective immunity against B cell lymphoma. Nat Med 2:1038–41

Tang D-C, DeVit M and Johnston SA (1992) Genetic immunization is a simple method for eliciting an immune response. Nature 356:152–154

Tao M-H and Levy R (1993) Idiotype/granulocyte-macrophage colony-stimulating factor fusion protein as a vaccine for B cell lymphoma. Nature 362:755–758

Tepper RI, Pattengale PK and Leder P (1989) Murine interleukin-4 displays potent anti-tumor activity in vivo. Cell 57:503–512

Townsend SE, Su FW, Atherton JM and Allison JP (1994) Specificity and longevity of antitumor immune responses induced by B7- transfected tumors. Cancer Res 54:6477–83

Trinchieri G (1995) Interleukin-12:a proinflammatory cytokine with immunoregulatory functions that bridge innate resistance and antigen-specific adaptive immunity. Annu Rev Immunol 13:251–76

Ullrich SJ, Robinson EA, Law LW, Willingham M and Appella E (1986) A mouse tumor-specific transplantation antigen is a heat shock-related protein. Proc Natl Acad Sci USA 83:3121–5

Ulmer JB, Donnelly JJ, Parker SE, Rhodes GH, Felgner PL, Dwarki VJ, Gromkowski SH, Deck RR, DeWitt CM, Friedman A, Hawe LA, Leander KR, Martinez D, Perry HC, Shiver JW, Montgomery DL and Liu M (1993) Heterologous protection against Influenza by injection of DNA encoding a viral protein. Science 259:1745–1749

Wakabayashi S, Taniguchi M, Tokuhisa T, Tomioka H and Okamoto S (1981) Cytotoxic T lymphocytes induced by syngeneic mouse melanoma cells recognize human melanomas. Nature 294:748–750

Wakimoto H, Abe J, Tsunoda R, Aoyagi M, Hirakawa K and Hamada H (1996) Intensified antitumor immunity by a cancer vaccine that produces granulocyte-macrophage colony-stimulating factor plus interleukin 4. Cancer Res 56:1828–33

Wang B, Boyer J, Srikantan V, Coney L, Carrano R, Phan C, Merva M, Dang K, Agadjanyan M, Gilbert L, Ugen K, Williams VW and Weiner DB (1993) DNA inoculation induces neutralizing immune responses against human immunodeficiency virus type 1 in mice and nonhuman primates. DNA Cell Biol 12:799–805

Wang B, Merva M, Dang K, Ugen KE, Williams WV and Weiner DB (1995) Immunization by direct DNA inoculation induces rejection of tumor cell challenge. Human Gene Therapy 6:407–418

Wang B, Ugen KE, Srikantan V, Agadjanyan MG, Dang K, Sato AI, Refaeli Y, Boyer J, Williams WV and Weiner DB (1993) Gene inoculation generates immune responses against human immunodeficiency virus type 1. Proc Natl Acad Sci USA 90:4156–4160

Williams WV, Boyer JD, Merva M, LiVolsi V, Wilson D, Wang B and Weiner DB (1993) Genetic infection induces protective in vivo immune responses. DNA Cell Biol 12:675–683

Wolff JA, Malone RW, Williams P, Chong W, Acsadi G, Jani A and Felgner PL (1990) Direct gene transfer into mouse muscle in vivo. Science 247:1465–1468

Xiang ZQ, Spitalnik SL, Cheng J, Erikson J, Wojczyk B and Ertl HC (1995) Immune responses to nucleic acid vaccines to rabies virus. Virology 209:569–79

DNA Vaccination as an Approach to Malaria Control: Current Status and Strategies

D.L. Doolan[1], R.C. Hedstrom[1], M.J. Gardner[1,2], M. Sedegah[1,2], H. Wang[1,3], R.A. Gramzinski[1], M. Margalith[4], P. Hobart[4], and S.L. Hoffman[1]

[1] Malaria Program, Naval Medical Research Institute, 12300 Washington Avenue, Rockville, MD 20852, USA

[2] Current address: The Institute for Genomic Research, 9712 Medical Center Drive, Rockville, MD 20850, USA

[3] Current address: Department of Parasitology, Union Medical College, Dong Dan 3 Tiao, Beijing 100005, China

[4] Vical Incorporated, 9373 Town Centre Drive, Suite 100, San Diego, CA 92121, USA

1 Introduction

During the twentieth century, the primary approach to malaria prevention has been to interfere with transmission of the parasite between the infected mosquito and the human host using physical barriers, insecticides and prophylactic drugs. Despite these measures, approximately 40% of the world's population is at risk and it is estimated that there are 300–500 million new *Plasmodium* infections and 1.5–2.7 million deaths annually due to malaria (WHO 1990). Accordingly, there have been major efforts to develop an efficacious malaria vaccine.

1.1 Life Cycle of *Plasmodium* spp.

When a person is infected with malaria by the bite of an infectious female anopheles mosquito, sporozoites in the salivary gland are inoculated into the peripheral circulation of the host and travel through the bloodstream to invade hepatocytes. This invasion must occur very rapidly since sporozoites cannot be detected in the circulation after 30 min of inoculation. Within the hepatocyte, uninucleate sporozoites undergo a cycle of asexual schizogony for a period of 2–10 days, depending on the species (a minimum of 5 days for human malarias), producing as many as 30000 uninucleate merozoites. The liver-stage parasite develops in a parasitophorous vacuole bounded by an extensive parasite membrane and host membrane. During this period of liver-stage maturation, there are no clinical symptoms of malaria. From each infected hepatocyte, thousands of merozoites are then released into the bloodstream to invade red blood cells, commencing the erythrocytic-stage of the disease. The parasite undergoes another asexual amplification, producing as many as 36 merozoites per schizont. After 48–72 h, the infected red blood cell ruptures, releasing merozoites which may invade other red blood cells. Alternatively, some differentiate into male and female gametocytes which undergo gametogenesis following ingestion by a mosquito, resulting in gametes that combine to form a zygote after exflagellation. The zygote differentiates further into an ookinete. This sexual reproduction occurs in the mosquito midgut and the resulting ookinete penetrates the midgut wall and develops into an oocyst, within which thousands of new sporozoites are produced. The sporozoites migrate to the salivary glands and can be inoculated into another host, thereby repeating the life cycle.

1.2 Rationale for the Development of a Pre-erythrocytic Malaria Vaccine

Malaria could be prevented if the invasion of sporozoites into hepatocytes could be blocked (by inducing protective antibodies that neutralize sporozoite infectivity for hepatocytes) or if parasites within the infected hepatocyte could be destroyed before rupture and liberation of merozoites (by protective T cell effector mechanisms, and perhaps antibodies). The development of a vaccine targeted at either the sporozoite in circulation or at the infected hepatocyte, collectively referred to as the pre-erythrocytic stage of the parasite's life cycle, has been the focus of our laboratory. An effective pre-erythrocytic stage vaccine would eliminate all infection before the appearance of clinical disease. Such a vaccine is feasible because immunization of rodents, primates and humans with radiation-attenuated *Plasmodium* sporozoites induces sterile protective immunity against malaria (reviewed in HOFFMAN et al. 1996). This was first established for humans in the early 1970s, but the irradiated sporozoite vaccine is not practical for large-scale human use and an effective malaria vaccine that induces protection comparable to that induced by the irradiated sporozoite vaccine is not available, although recent data are encouraging (STOUTE et al. 1997). During the past decade, there has however been significant progress towards identifying the protective immune mechanisms, identifying the antigenic targets of these immune mechanisms, and developing vaccine delivery systems that induce the desired immune response against the desired target (reviewed in HOFFMAN et al. 1996).

There is now a large body of evidence implicating $CD8^+$ T cells as critical for sporozoite immunity. Protection is dependent on $CD8^+$ T cells in some rodent models (SCHOFIELD et al. 1987; WEISS et al. 1988) and $CD8^+$ T cells can protect in the absence of other parasite-specific immune responses (ROMERO et al. 1989; RODRIGUES et al. 1991; WEISS et al. 1992; KHUSMITH et al. 1994). Immunity is presumably mediated via $CD8^+$ T cell recognition of parasite-derived peptides that are presented on the surface of infected hepatocytes in association with class I MHC molecules. It has been demonstrated that $CD8^+$ cytotoxic T lymphocytes (CTLs) can recognize an epitope from the *Plasmodium yoelii* circumsporozoite protein (PyCSP) expressed on *P. yoelii* infected hepatocytes (WEISS et al. 1990). That MHC-restricted immune-mediated elimination of infected hepatocytes which are expressing parasite antigens can protect against sporozoite-transmitted malaria has also been demonstrated (HOFFMAN et al. 1989, 1990). Accordingly, there have been efforts to develop vaccines that induce protective $CD8^+$ T cells.

1.3 Conventional Vaccines vs DNA Vaccines

The use of conventional vaccine delivery systems such as synthetic peptides, purified recombinant proteins and live recombinant organisms has not induced adequate protective immunity against malaria. This is best illustrated in the highly infectious *P. yoelii* model system (ID_{50} of 1–2 sporozoites; KHUSMITH et al. 1991, 1994), with reference to the PyCSP which has been identified as one target antigen of protective $CD8^+$ T cells (reviewed in HOFFMAN et al. 1996). There was no protection after immunization with recombinant vaccinia (SEDEGAH et al. 1988), *Salmonella typhimurium* (SEDEGAH et al. 1990), or pseudorabies (FLYNN et al. 1990; SEDEGAH et al. 1992) expressing PyCSP, despite high levels of CSP-specific CTL activity (FLYNN et al. 1990; SEDEGAH et al. 1992). $CD8^+$ T cell-dependent protection was achieved after immunization with recombinant P815 mastocytoma cells expressing PyCSP (KHUSMITH et al. 1991) and with a recombinant influenza virus/vaccinia virus combination (Li et al. 1993; Rodrigues et al. 1994). However mastocytoma cells cannot be administered to humans and the use of live-attenuated vaccines has many disadvantages (summarized in Hedstrom et al. 1994).

In contrast, the recent technology of DNA vaccines looks promising. DNA immunization involves the direct introduction of a plasmid DNA encoding a target antigen which is subsequently expressed within cells of the host and can induce an immune response. The simplicity of the DNA approach implies that it should be possible to combine many DNAs, each encoding different antigens, and thereby broaden the immune response. This would be particularly relevant to malaria since it is well established that there are few target epitopes (at least for T cells), that these are often variant, and that the specific variants often do not cross-react immunologically. Thus the administration of multiple antigens may be required to induce effective immunity (reviewed in Doolan and Hoffman 1997).

1.4 Approach to the Development of a DNA Vaccine Against Malaria

Our strategy for the development of a DNA vaccine designed to protect humans against falciparum malaria is to first demonstrate protective efficacy and biological activity of the induced immune responses in the *P. yoelii* rodent model with *P. yoelii* genes that encode protective antigens, the homologs of which have been identified in the *P. falciparum* parasite. The major focus is on pre-erythrocytic antigens, although erythrocytic-stage antigens are also being studied. The ultimate *P. falciparum* combination DNA vaccine would be comprised of a component designed to induce T cell responses against pre-erythrocytic-stage antigens expressed in the infected hepatocyte, together with another component designed to induce protective antibodies and perhaps $CD4^+$ T cells against erythrocytic-stage antigens as a preventative measure (Doolan and Hoffman 1997).

DNA vaccines encoding pre-erythrocytic stage antigens are evaluated in terms of: (1) capacity to induce $CD8^+$ and $CD4^+$ T cell and B cell responses, and cytokine production; (2) capacity to protect against sporozoite and blood-stage challenge; and (3) ability of induced antibodies to inhibit sporozoite invasion and liver-stage development. DNA vaccines encoding erythrocytic-stage antigens are evaluated in terms of: (1) capacity to induce B and T cell responses; (2) capacity to protect against sporozoite and blood-stage challenge; and (3) ability of induced antibodies to inhibit the growth of *Plasmodium* spp. in vitro and passively transfer protection to naive mice in vivo. In the *P. yoelii* system, the PyCSP, *P. yoelii* hepatocyte erythrocyte protein 17 (PyHEP17) and the *P. yoelii* sporozoite surface protein 2 (PySSP2) have been employed as model pre-erythrocytic-stage antigens, and the *P. yoelii* merozoite surface protein-1 (PyMSP-1) as a model erythrocytic-stage antigen.

Aotus monkeys are being studied in parallel to assess the immunogenicity of selected DNA vaccines as a prelude to assessing protective efficacy of *P. falciparum* pre-erythrocytic and erythrocytic-stage combination DNA vaccines in human clinical trials. The immunogenicity of these vaccines with respect to the induction of cellular immune responses is being analyzed in *Rhesus* nonhuman primates.

A number of strategies designed to optimize the protective immune responses to individual antigens in preparation for the formulation of a combination DNA vaccine are also being investigated.

2 *Plasmodium yoelii* Rodent Model

2.1 Immunization Regime

It is well established that both MHC and non-MHC genes control protection against *P. yoelii* in mice (WEISS et al. 1989) and that immune responses to many *P. falciparum*, *P. vivax*, *P. yoelii* and *P. berghei* $CD8^+$ and $CD4^+$ T cell epitopes and B cell epitopes are genetically restricted (GOOD 1994). Therefore, to investigate DNA-induced protective efficacy in the *P. yoelii* rodent model, five inbred mouse strains differing in both genetic background and H-2 haplotype were studied: BALB/cByJ (H-2^d); A/J (H-2^a); B10.BR/SgSnJ, (H-2^k); B10.Q/SgJ (H-2^q); and C57BL/6J (H-2^b) (The Jackson Laboratory, Bar Harbor, ME). Outbred CD-1 mice (Charles River Laboratories, Wilmington, MA) were also studied since they are considered more representative of an outbred human population.

In the standard immunization regimen, mice were immunized three times at 3 week intervals intramuscularly in each tibialis anterior muscle with 50 μg DNA in 50 μl saline. Mice were challenged 2 weeks after the third immunization by tail vein injection with 100 infectious *P. yoelii* (17X NL, clone 1.1) sporozoites or 200 parasitized erythrocytes. The ID_{50} for *P. yoelii* in mice is 1–2 sporozoites (KHUSMITH et al. 1991, 1994; M SEDEGAH and DL DOOLAN, unpublished data). Giemsa-stained blood smears were examined on days 5–14. Protection was defined as the complete absence of blood-stage parasitemia.

2.2 *Plasmodium yoelii* Circumsporozoite Protein (PyCSP)

The PyCSP (391 amino acids; LAL et al. 1987) migrates on reducing SDS-PAGE higher than predicted as two bands of 75 kDa and 56 kDa. The antigen is expressed on the surface of sporozoites and on the parasite membrane and parasitophorous vacuole membrane of developing exoerythrocytic parasites. It is a target of protective immune responses. In vivo, PyCSP-specific $CD8^+$ (ROMERO et al. 1989; RODRIGUES et al. 1991; WEISS et al. 1992) and $CD4^+$ (RENIA et al. 1993) T cell clones adoptively transferred protection. In vitro, PyCSP-specific $CD8^+$ CTLs eliminated infected hepatocytes from culture in an MHC-restricted manner (WEISS et al. 1988).

Since the PyCSP is a target of protective immune responses, two plasmids encoding the PyCSP gene were constructed. The pDIP and nkCMVintpolyli vectors have been described previously (SEDEGAH et al. 1994b). In both vectors, expression of the encoded gene is driven by a cytomeglavirus (CMV) immediate/early gene promoter. The nkCMVint vector also contains the intron A of the immediate/early gene of CMV, reported to positively regulate expression from the CMV immediate/early enhancer/promoter (CHAPMAN et al. 1991). In the pDIP vector, the inserted gene is fused in-frame with the first 82 amino acids of human interleukin-2 (CULLEN 1986). In vitro expression of the PyCSP antigen encoded by these constructs was confirmed by transfection of various cultured cells and immunoblot analysis of cell lysates (RC HEDSTROM and M MARGALITH, unpublished data). Initial studies showed that 56% (9/16) of BALB/c mice immunized with pDIP/PyCSP DNA were protected following sporozoite challenge (SEDEGAH et al. 1994b). In additional studies, protection as high as 83% has been achieved in BALB/c mice following immunization with nkCMVint/PyCSP DNA (M SEDEGAH and DL DOOLAN, unpublished data).

PyCSP DNA immunization of BALB/c mice induced high levels of CSP-specific antibodies in a dose-dependent manner: the higher the dose of DNA, the greater the antibody response, as assessed by both indirect fluorescent antibody test (IFAT) against air-dried *P. yoelii* sporozoites and by ELISA against recombinant protein (SEDEGAH et al. 1994b). Antibody titers were 10–15 times higher than titers induced by immunization with irradiated sporozoites. However, the DNA-induced antibodies had little biological activity, as assessed by the in vitro inhibition of sporozoite invasion and development assay. Presumably the antibodies were not focused on the primary protective epitope in the repeat region, but also recognized epitopes in non-repeat regions (SEDEGAH et al. 1994b).

Immunization with PyCSP DNA also induced antigen-specific, MHC-restricted $CD8^+$ CTLs directed against a previously identified nonamer CTL epitope on the PyCSP (SEDEGAH et al. 1994b). The presence of CTLs correlated with the presence of antibodies. Interestingly, the levels of CTLs induced by PyCSP DNA immunization (70%–80% specific lysis) were significantly greater than the levels induced by protective immunization with irradiated sporozoites (20%–30% specific lysis). In addition, there was a 86% reduction in the number of liver-stage schizonts following massive sporozoite challenge in the PyCSP DNA immunized mice, as compared with mice immunized with control plasmid without any inserted genes (SEDEGAH et al. 1994b).

The protective immunity induced by immunization with PyCSP DNA was genetically restricted (DOOLAN et al. 1996a). BALB/c (H-2^d) were highly protected against sporozoite challenge but four other inbred mouse strains were not protected. The antibody response was also genetically restricted: three strains of mice were high responders and two strains were low responders (DOOLAN et al. 1996a). These data predict that a vaccine based on the CSP alone will be inadequate to protect an outbred human population and support the concept of a multivalent malaria vaccine (DOOLAN and HOFFMAN 1997).

2.3 *Plasmodium yoelii* Hepatocyte Erythrocyte Protein 17 (PyHEP17)

PyHEP17 is a 17 kDa protein identified on the parasitophorous vacuole membrane of *P. yoelii*-infected hepatocytes and erythrocytes (CHAROENVIT et al. 1995). It is recognized by an IgG1 monoclonal antibody NYLS3. In vitro, NYLS3 eliminated liver-stage parasites from culture in an antigen-specific manner and in vivo NYLS3 reduced the density and delayed the onset of blood-stage parasitemia (CHAROENVIT et al. 1995). This suggested that PyHEP17 may be an important target of protective immune responses.

PyHEP17 is composed of two exons (DOOLAN et al. 1996b). A fragment of the PyHEP17 cDNA consisting of the complete exon 1 and 57% of exon 2 but not including the NYLS3 epitope (PyHEP17Ex1.2), was cloned into the nkCMVint vector. In vitro expression of the antigen in this construct was confirmed by IFAT and RT-PCR of transfected COS cells (DL DOOLAN, unpublished data).

Immunization with PyHEP17Ex1.2 DNA protected three of five inbred strains (A/J, BALB/c and B10.BR) against sporozoite but not blood-stage challenge (DOOLAN et al. 1996a). Additional partial and full-length constructs of PyHEP17 which included the NYLS3 epitope also conferred protective immunity (DOOLAN et al. 1996b; DL DOOLAN, unpublished data).

Antigen-specific MHC-restricted $CD8^+$ CTLs were induced in BALB/c (H-2^d) mice by immunization with PyHEP17 DNA (DOOLAN et al. 1996b; DL DOOLAN, unpublished data). No CTLs were detected in PyHEP17 DNA immunized A/J nor B10.BR mice but this may reflect a problem with the assay system.

In contrast to the PyCSP, the presence of PyHEP17 CTL did not correlate with antibody (DL DOOLAN, unpublished data). Indeed, PyHEP17 DNA immunization induced little or no antibody response in all five strains tested. Sera did not recognize *P. yoelii* liver-stage parasites and recognized *P. yoelii* blood-stage parasites by IFAT only at very low levels (DOOLAN et al. 1996a). Furthermore, there was no correlation between antibody titers and protection; some mice which were protected had no detectable antibodies and the highest antibody titers were observed in C57BL/6 mice which were not protected against sporozoite challenge (DOOLAN et al. 1996a).

2.4 PyHEP17/PyCSP Combination DNA Vaccine

The protective immunity induced by immunization with either PyHEP17 or PyCSP DNA was genetically restricted (DOOLAN et al. 1996a). However, bivalent DNA immunization was found to circumvent the genetic restriction of protection found after immunization with the PyHEP17 or PyCSP DNA vaccine delivered alone. Immunization of mice with a combination of PyHEP17 and PyCSP DNA protected 80%–90% of both high responder strains (A/J and BALB/c) as well as the moderate responder strain (B10.BR) (DOOLAN et al. 1996a). In the two nonresponder strains (B10.Q and C57BL/6), a delay in the onset of parasitemia of up to 4 days was noted in most mice with the combination relative to either vaccine alone.

consistent with an effect of the combination at the liver-stage (DL Doolan, unpublished data). Therefore, bivalent DNA immunization protected mice of diverse genetic backgrounds and H-2 haplotypes. Furthermore, the data with the moderate responder strain suggested that this protection was additive.

2.5 *Plasmodium yoelii* Sporozoite Surface protein 2 (PySSP2)

PySSP2 is an 826 amino acid antigen, but migrates on reducing SDS-PAGE at 140 kDa (Rogers et al. 1992). It is expressed on the surface of sporozoites and on the parasite membrane of developing exo-erythrocytic parasites. PySSP2 is a target of protective immune responses. In vivo, PySSP2-specific $CD8^+$ T cell clones adoptively transferred protection (Khusmith et al. 1994). In vivo, immunization with mastocytoma cells expressing a combination of PyCSP and PySSP2 conferred $CD8^+$ T cell-dependent protective immunity (Khusmith et al. 1991).

The full-length PySSP2 gene was cloned into the nkCMVint vector, in a number of different constructions (H Wang, unpublished data): (1) nkCMVint/PySSP2, full-length PySSP2; (2) nkCMVint/PySSP2r, a more defined construct containing only the repeat region of PySSP2, known to be the target of protective $CD4^+$ T cells (Wang et al. 1996); (3) nkCMVint/PySSP2:cs, a chimeric PySSP2 construct in which the native transmembrane and cytoplasmic domains were replaced with the anchor domain of PyCSP. In vitro expression of the antigen in these constructs was confirmed by IFAT and immunoblot analysis of cell lysates. Immunization with each of the constructs nkCMVint/PySSP2, nkCMVint/PySSP2r, and nkCMVint/PySSP2:cs conferred protection against sporozoite challenge in A/J mice (H Wang, unpublished data). The unmodified full-length gene was the most protective.

PySSP2 DNA immunization of three of four strains of inbred mice with each of the three constructs induced moderate to high levels of SSP2-specific antibodies, as assessed by both IFAT against air-dried *P. yoelii* sporozoites and by ELISA against synthetic peptide $(NPNEPS)_3$ (H Wang, unpublished data).

2.6 *Plasmodium yoelii* Merozoite Surface Protein-1 (PyMSP-1)

PyMSP-1 is a 230 kDa protein (1772 amino acids) (Farley et al. 1994; Lewis 1989) synthesized during the development of the schizont and expressed on the surface of late-stage schizonts and merozoites (reviewed in Diggs et al. 1993). In vivo data demonstrate that MSP-1 is a target of protective immune responses. Passive immunization with monoclonal antibodies to *P. falciparum* and *P. yoelii* MSP-1 reduced the efficiency of erythrocyte invasion in humans and mice, respectively (Burns et al. 1989; Blackman et al. 1990). Active immunization with *P. falciparum* MSP-1 protected *Aotus* (Siddiqui et al. 1987) and *Saimiri* (Perrin et al. 1984; Etlinger et al. 1991) monkeys against blood-stage challenge. Immunization with a recombinant protein of *P yoelii* MSP-1 (Holder and Freeman 1981; Freeman and HOLDER 1983) or with recombinant forms of the 15 kDa COOH-terminal portion of PyMSP-1 (DALY and LONG 1993, 1995) protected mice. This protection was antibody-mediated. Antibodies recognized a conformational epitope, since protection was abolished upon reduction and alkylation of the COOH-terminal protein (LING et al. 1994). The COOH-terminal of PyMSP-1 is therefore being studied as a model antigen to determine whether immunization with plasmid DNA encoding malarial genes can induce protective antibodies. Cellular immunity has, however, also been implicated in MSP-1 induced protection (reviewed in Diggs et al. 1993). Recent data suggests that $CD4^+$ T cells may be required to completely eliminate infection (DALY and LONG 1995).

Three plasmids encoding the PyMSP-1 COOH-terminal were constructed (MJ GARDNER, unpublished data): (1) nkCMVint/PyMSP-1, PyMSP-1 COOH-terminal cloned into the nkCMVint vector; (2) nkCMVint-TPA-P2P30/PyMSP-1, PyMSP-1 COOH-terminal fused with two promiscuous T helper epitopes derived from tetanus toxin, P2 and P30 (PANINA BORDIGNON et al. 1989; VALMORI et al. 1992) and with the signal peptide of tissue plasminogen activator protein (TPA); and (3) pDIP/MSP-1, PyMSP-1 COOH-terminal cloned into the pDIP vector. BALB/c mice were immunized intradermally with each of these constructs and challenged with 200 parasitized erythrocytes. After challenge, three out of ten mice immunized with pDIP/PyMSP-1 and one out of ten mice immunized with nkCMVint-TPA-P2P30/PyMSP-1 exhibited a delay in patency, a slower rise to peak parasitemia, and extended survival as compared to controls (MJ GARDNER, unpublished data). A statistically significant difference in geometric mean parasitemia was detected over days 5–8 between mice immunized with any of the PyMSP-1 constructs and mice immunized with control plasmids.

Immunization with pDIP/PyMSP-1 and with nkCMVint-TPA-P2P30/PyMSP-1 induced significant levels of PyMSP-1-specific antibodies, as assessed by both IFAT against air-dried *P. yoelii* infected erythrocytes and by ELISA against recombinant protein (M.J. GARDNER, unpublished data). However, induced antibody responses were not comparable to the titers induced by immunization with recombinant protein in adjuvant (MJ GARDNER, unpublished data). Immunization with nkCMVint/PyMSP-1 failed to induce significant antibodies.

2.7 Summary

1. Immunization with PyCSP DNA protected one out of five strains of inbred mice (BALB/c) against sporozoite challenge.
2. Immunization with PyHEP17 DNA protected three out of five strains of inbred mice (A/J, BALB/c and B10.BR) against sporozoite but not blood-stage challenge.
3. Immunization with PySSP2 DNA protected one strain of inbred mice (A/J) against sporozoite challenge.
4. Immunization with PyMSP-1 COOH-terminal DNA partially protected one strain of inbred mice (BALB/c) against blood-stage challenge.

5. Immunization with PyHEP17-PyCSP combination protected three out of five strains of inbred mice (A/J, BALB/c and B10.BR) $>80\%$ against sporozoite challenge, and affected liver-stage parasite development in two other strains (B10.Q and C57BL/6).
6. Bi-gene DNA immunization circumvented the genetic restriction of protection to the individual components of the vaccine.
7. Immunization with PyCSP DNA and with PyHEP17 DNA induced antigen-specific MHC-restricted $CD8^+$ CTLs.
8. The $CD8^+$ CTL response induced by PyCSP DNA immunization was of a much greater magnitude than the response induced by protective immunization with irradiated sporozoites.
9. Immunization with PyCSP DNA induced high titers of antigen-specific antibodies. These antibodies had little biological activity.
10. Immunization with PyHEP17 DNA induced little or no antibody response.
11. Immunization with PySSP2 DNA induced moderate titers of antigen-specific antibodies.
12. Immunization with PyMSP-1 COOH-terminal DNA induced significant titers of antigen-specific antibodies.
13. Immunization with plasmid DNA may induce antibodies which recognize conformational as well as linear B cell epitopes.

3 *Aotus* Primate Model

The *Aotus lemurinus lemurinus* monkey is a model system (Collins 1994) for the development of DNA vaccines designed to produce protective antibodies against *P. falciparum*. Unlike other nonhuman primates such as *Rhesus* monkeys and chimpanzees, *Aotus* monkeys can develop blood-stage parasitemia following challenge with sporozoites of some *P. falciparum* strains and thus can be used for the assessment of protective efficacy conferred by candidate anti-malarial vaccines, as well as in the development of anti-malarial drugs.

3.1 *Plasmodium yoelii* CSP

Initially, the immunogenicity of *Plasmodium* spp. antigens administered as plasmid DNA was modeled in the *Aotus* monkey using PyCSP. Monkeys were immunized either intradermally or intramuscularly in each tibialis anterior muscle with or without pre-treatment with bupivacaine. Bupivacaine is myotoxic when injected into skeletal muscle and may therefore result in enhanced plasmid expression during muscle regeneration (Wang et al. 1993; Danko et al. 1994). Three doses of 125, 500 or 2000 μg plasmid DNA were administered at 3 week intervals and monkeys were boosted at week 47. A multiple antigen peptide construct (MAP)

(WANG et al. 1995) of the dominant PyCSP B cell epitope, QGPGAP, linked to the two promiscuous T helper epitopes derived from tetanus toxin, P2 and P30 (PANINA BORDIGNON et al. 1989; VALMORI et al. 1992), was used as a positive control. This construct induced very high antibody titers in the rodent model and completely protected against sporozoite challenge (WANG et al. 1995).

Immunization of monkeys intramuscularly with or without bupivacaine pretreatment failed to induce significant levels of antibodies. However, intradermal immunization of monkeys induced high titered antibodies against *P. yoelii* sporozoites, as assessed by both the IFAT against air-dried sporozoites and ELISA against recombinant protein (GRAMZINSKI et al. 1996). As seen with the PyCSP DNA in BALB/c mice, there was a general dose response: the higher the dose of DNA, the greater the antibody response. After the first three immunizations, antibody titers peaked briefly at week 9 and declined to 50% of their values by week 14, and 2%–20% by week 46. Induced antibody responses were not comparable to the titers induced by the MAP vaccine. Two weeks following the booster immunization at week 47, however, anti-CSP antibody titers were markedly elevated and were equivalent in titer to antibodies generated by the MAP vaccine (GRAMZINSKI et al. 1996).

3.2 Summary

1 Intradermal but not intramuscular immunization of *Aotus* monkeys with PyCSP DNA induced high antibody responses equivalent to those induced by a MAP/adjuvant vaccine.
2. A rest period between immunizations may result in elevated antibody titers.
3. DNA vaccines may induce protective antibodies against *P. falciparum* antigens in *Aotus* monkeys.

4 Strategies for the Optimization of Protective Immune Responses

Our rationale is to optimize the immune response to individual DNA vaccine components as a starting point for the formulation of a combination DNA vaccine. To achieve this, we are examining several strategies.

4.1 Optimization of Dose Regimen and Immunization Schedule

Studies with PyCSP, PyHEP17, PySSP2 and the PyHEP17/PyCSP combination suggest that each DNA vaccine may require distinct formulation and administration schemes for optimization of protective efficacy. Thus, we have been investigating the route of immunization, dose, dosing interval and number of doses.

4.1.1 Route vs Protection

With PyHEP17, comparable levels of protection were induced by both intramuscular and intradermal immunization (DL Doolan, unpublished data). With PyCSP and PySSP2, protection was induced by intramuscular immunization but little or no protection was induced by intradermal immunization (M. Sedegah and H. Wang, unpublished data).

4.1.2 Route vs Cytotoxic T Lymphocytes

With PyCSP, significantly higher levels of CTLs were induced by intramuscular compared with intradermal immunization (Sedegah et al., submitted). This may explain the difference in protection conferred by the different routes of immunization.

4.1.3 Route vs Antibodies

With PySSP2, intradermal immunization induced higher antibody titers than did intramuscular immunization (H Wang, unpublished data). With PyHEP17 and PyCSP, the route of immunization had little or no effect on antibody titer (DL Doolan and M Sedegah, unpublished data).

Conventionally, MHC class I restricted T cells recognize peptides derived from processing of endogenously synthesized proteins, whereas class II restricted T cells recognize peptide fragments derived from exogenous proteins presented by antigen presenting cells. Preliminary data in our system suggest that the intramuscular route of immunization may be optimal for induction of CTLs and $CD8^+$ T cell mediated immune responses, but that the intradermal route may be optimal for induction of antibody responses. This dichotomy may reflect the accessibility of the target antigen to the endogenous and exogenous pathways of processing and presentation, respectively.

4.1.4 Dose vs Antibody

With PyCSP, in both mice and monkeys, there was a dose response in antibody production – the more DNA administered, the higher the antibody titer (M. Sedegah, manuscript submitted; Gramzinski et al. 1996). In contrast, preliminary results with PySSP2 indicated that immunization with 50 μg DNA induced significantly higher antibody titers compared with 100 or 200 μg doses (M Sedegah, unpublished data).

4.1.5 Number of Doses vs Protection

With both PyHEP17 and PyCSP, two or three immunizations at 2 or 3 week intervals conferred protection against sporozoite challenge, but a fourth immunization completely abrogated protection (DL Doolan and M Sedegah, unpublished data). Little or no protection was conferred by a single immunization (DL Doolan and M Sedegah, unpublished data) despite the induction of high levels of $CD8^+$ CTLs (M Sedegah, unpublished data).

We speculate that this anomaly may reflect a switch from a Th1 to Th2 type cytokine profile, as has been demonstrated with *Schistosoma mansoni*. (Pearce et al. 1991). Consistent with this, it has been shown that the protection conferred by immunization with PyHEP17 DNA and with the PyHEP17/PyCSP DNA combination is absolutely dependent on both $CD8^+$ T cells and interferon-γ, a Th1 type cytokine (Doolan et al. 1996a).

4.1.6 Immunization Regime

Preliminary data suggested that with PyHEP17, immunization at 2 week intervals conferred greater protection than immunization at 3 week intervals (DL Doolan, unpublished data). In contrast, with PyCSP, increasing the interval between immunizations from 3 weeks to 6 weeks may have improved the protection (M Sedegah and L Scheller, unpublished data). However, in these experiments, numbers were small and the differences were not significant.

4.2 Modification of the Plasmid Vector or Encoding Gene Sequence to Alter Presentation or Improve Mammalian Expression

The malarial genes used in the studies discussed here each are predicted to encode the native signal peptide which may be differentially recognized and/or processed within mammalian cells. Standardization of the plasmid constructs with regard to mammalian processing signals may aid consistency of induced immune responses, and modification of the constructs may increase vaccine efficacy. A number of strategies are therefore under consideration.

4.2.1 Human Use Vectors with Improved Expression

The plasmid vectors are ultimately intended for human use. Therefore, extraneous sequences, particularly those derived from SV40, have been removed from those vectors used in initial studies. This has resulted in the introduction of a "human use" vector, designated VR1012 (Hartikka et al. 1996). In this vector, the expression of the inserted gene is driven by the CMV promoter/intron A sequence and termination sequences are derived from the bovine growth hormone gene. The VR1012 vector expresses encoded reporter proteins at levels 10–100 times higher than those achieved with the nkCMVint plasmid (P Hobart, unpublished data). A second human use vector, designated VR1020, was constructed by ligating a DNA fragment from the human tissue plasminogen activator (hTPA) gene, including the leader peptide (residues 1–23), into the VR1012 backbone (Luke et al. 1997). Genes are cloned in-frame with the hTPA leader sequence and expressed proteins are thought to be preferentially targeted into the secretory pathway.

4.2.2 Protein Localization

The presence or absence of signal peptide sequences may influence the elicited immune response, by directing whether the target antigen is presented outside the cells, by secretion or surface presentation, or remains in the cytoplasm of the transfected host cell. A nonsecreted antigen may be optimal for $CD8^+$ T cell mediated protection, since intracellular expression may be preferential for presentation via the endogenous class I pathway. Alternatively, a secreted antigen might preferentially stimulate T helper cellular and humoral immune responses by preferentially directing processing and presentation of antigen through the endosomal pathway and facilitating uptake by professional antigen presenting cells; secretion may therefore be optimal for antibody-mediated protection. Indeed, sustained antibody levels may be achieved if the transfected host cell acts as an antigen reservoir, releasing small amounts of antigen for a prolonged time. By designing plasmid vectors that do or do not encode secretory signals for either secretion of the target antigen into the extracellular space or retention of the antigen intracellularly, the predominant type of immunity elicited by DNA immunization may be manipulated to enhance protection. Derivatives of the nkCMVint and VR1012 vectors which contain the first 78 amino acids of TPA protein, including the TPA secretory signal sequence, have been developed to maximize secretion of encoded genes. Genes are inserted in-frame with the TPA signal sequence of the vector and may also have the anchor domain removed to prevent retention of the protein on the cell surface.

In the *P. yoelii* rodent model, the PySSP2 gene was used as a model antigen to assess increase in immunogenicity associated with modification of the plasmid construct (RC Hedstrom, unpublished data). The ability of plasmid constructs to express PySSP2 was evaluated in vitro by transient transfection of a melanoma cell line followed by immunoblot analysis of the cell lysate. In both the nkCMVint and the VR1012 vector, addition of the TPA signal appeared to markedly enhance the expression of the inserted gene (RC Hedstrom, unpublished data). Furthermore, the level of expression of PySSP2 in the VR1012 construct was much greater than that in the nkCMVint vector. An immunoreactive band of the appropriate size was detected in tissue culture supernatants from transfections of the TPA signal fusion constructs, consistent with secretion of the antigen, but not from transfections of the constructs lacking the signal sequence (R.C. Hedstrom, unpublished data). These PySSP2 constructs have not yet been fully evaluated In vivo.

4.2.3 Modification of the Encoded Gene

Preliminary results from our laboratory suggest that there may be specific but as yet undefined features of some malarial genes that determine the degree of expression in vivo, the result of which is to alter the immunogenicity of the encoded antigen. For instance, immunization with nkCMVint/PyCSP DNA induced $CD8^+$ T cell-dependent protection and high antibody titers (Sedegah et al. 1994b), but nkCMVint/PyHEP17 DNA immunization induced $CD8^+$ T cell-dependent protection and little or no antibody response (DOOLAN et al. 1996a). We therefore investigated whether the immunogenicity of PyHEP17 could be increased by inserting the PyHEP17 cDNA within the coding region of the PyCSP construct (replacing the *Hinc*II fragment, comprising 75% of the gene). Antibody responses induced by this construct in each of four mouse strains [(BALB/c (H-2^d), A/J (H-2^a), B10.BR (H-2^a) and C57BL/6 H-2^b)] were four-fold higher than the responses induced by the original PyHEP17 DNA vaccine, as assessed by IFAT against air-dried *P. yoelii* blood-stage parasites (RC HEDSTROM and DL DOOLAN, unpublished data).

4.2.4 Protein Expression

The differential activities of viral promoters in host cells may be useful for increasing vaccine efficacy. Differences in the levels and persistence of reporter gene expression have been noted between CMV promoter-driven expression and Rous sarcoma virus (RSV) promoter-driven expression in muscle cells. It has been suggested that the initial higher level expression associated with the CMV promoter may be down-regulated in muscle cells, whereas RSV promoter associated expression may not be down-regulated and may therefore persist for a longer period of time (MANTHORPE et al. 1993). DNA vaccines could therefore be formulated utilizing both CMV and RSV promoter-driven expression so that initial high levels of CMV promoter-driven expression could induce vigorous anti-target immunity, and subsequent lower level but durable RSV promoter-driven expression of the target antigen could maintain the immunity. The concept of a single-dose vaccine might be approached in this manner.

4.2.5 Tissue Targeting

DNA vaccines that target specific tissues for presentation have the potential to enhance immune responses. In malaria, hepatic cell presentation of antigen is thought to be necessary for the induction of protective pre-erythrocytic immunity. Preferential delivery of pre-erythrocytic DNA vaccines to the liver may be accomplished with a poly (L-lysine) DNA vaccine complex by, for example, intravenous injection of a ligand for the asialoglycoprotein receptor in liver hepatocytes complexed to plasmid DNA encoding the target protein human factor IX (PERALES et al. 1994).

4.2.6 T Cell Costimulation

It is thought that antigen-specific T cell activation requires both engagement of the T cell receptor/CD3 complex to cognate antigen/MHC class I/II on antigen presenting cells and delivery of a second costimulatory signal to the T cell. Activation of the T cell surface receptor CD28 appears to deliver the dominant co-stimulatory signal during T cell activation (reviewed in LEE et al. 1991); the ligand(s) for CD28 are the B7 family of genes normally found on professional antigen presenting cells.

CD28 activation appears to play an important part in the generation of both cellular and humoral immune responses. By designing DNA vectors that simultaneously express both malarial genes and ligands capable of costimulating T cells, such as B7.1 or B7.2, it may be possible to manipulate this activation and thereby direct and/or augment the induced immune response.

4.2.7 Response Modification

Cytokines play a central role in the effector phase of the immune response in vivo and can effectively manipulate the immune response when administered experimentally. The immune response to a DNA vaccine may therefore be manipulated or augmented to favor Th1- vs Th2-type immune responses (Raz et al. 1993; Xiang and Ertl 1995) by co-administering particular combinations of immunomodulatory cytokine genes, e.g., interferon (IFN)-γ, granulocyte/macrophage colony-stimulating factor (GM-CSF), interleukin (IL)-12, and IL-1β or by simultaneously expressing both malarial genes and cytokine genes. In the case of malaria, this strategy could be modeled using the IL-12 gene, since preadministration with IL-12 protein completely protected mice (Sedegah et al. 1994a) and monkeys (Hoffman et al. 1997) against sporozoite challenge.

4.2.8 Minigene DNA Vaccines

Our approach to the design of DNA vaccines has been to utilize full-length genes in an attempt to include all T cell and B cell epitopes and thereby minimize the problem of genetic restriction to target epitopes. However, synthetic minigenes encoding known protective T cell and B cell epitopes, rather than entire antigens, are also being constructed as DNA vaccines in an effort to focus the immune response on known protective epitopes and avoid diversion of the induced immune response to nonprotective regions. The rationale for this strategy is the observation that antibodies directed to the repeat region of PyCSP are protective (Charoenvit et al. 1991), but antibodies directed to non-repeat regions are not protective (Sedegah et al. 1994b). Minigene epitopes may be derived from the same or different antigens, from one or more stages of the parasite's life cycle. Epitopes may be linked to promiscuous T helper epitopes which bind to multiple class II MHC molecules to provide T cell help for antibody production or cellular immune responses. Poor responses to conventional subunit vaccines may be attributed to failure to provide help, either because T helper epitopes were not present or, if present, were not recognized in the context of the MHC class II alleles of the individual. Ideally, the T helper epitope would be derived from a malarial antigen which would allow for boosting of immune responses following natural exposure to sporozoites. Work is underway to identify such an epitope.

Preliminary data in mice indicated that a minigene vaccine based on the immunodominant repeat of the PyCSP (QGPAP) induced moderate titers of anti-repeat antibodies (MJ Gardner, unpublished data) and a minigene vaccine comprising four repeats of the PySSP2 repeat B cell epitope (NPNEPS) conferred some protective immunity (R WANG, unpublished data).

4.2.9 Summary

1. Expression of the encoded antigen of a DNA vaccine can be significantly enhanced by choice of plasmid and alterations of the gene sequence.
2. Antigenicity of the encoded antigen can be manipulated by altering the gene sequence to include mammalian signaling sequences or to remove native processing signals.
3. Control of expression levels may be necessary for optimal protection.
4. Optimal induction of immunogenicity and protective efficacy by immunization with plasmid DNA may be antigen-dependent.

5 Conclusions

There will undoubtedly be problems in regard to the design and construction of human malaria vaccines that induce the required protective immune responses against the desired target. Nevertheless, for the first time a single vaccine delivery system that can be used in humans, a DNA vaccine, has induced a protective $CD8^+$ T cell response comparable to or perhaps even superior to that induced by irradiated sporozoites, and a bi-gene DNA vaccine has led to circumvention of genetic restriction of protection to its components. Furthermore, as well as protective $CD8^+$ T cell responses, this method of immunization has now reproducibly been shown to induce potent $CD4^+$ T cell-dependent antibody responses against both linear and conformational epitopes. The studies discussed here lay the foundation for multi-gene *P. falciparum* DNA vaccines designed to protect humans against this complex, intracellular parasite.

Acknowledgements. We thank S. Matheny, R. Lim and M. Chaisson for providing the *P. yoelii* sporozoites, M. Margalith for purification of nkCMVint*PyCSP*, and K. Gowda for technical assistance. The experiments reported herein were conducted according to the principles set forth in the "Guide for the Care and Use of Laboratory Animals," Institute of Laboratory Animal Resources, National Research Council, National Academy Press, 1996. The opinions and assertions herein are those of the authors and are not to be construed as official or as reflecting the views of the U.S. Navy or naval services at large. This work was supported by the Naval Medical Research and Development Command work units 62787A00101EVX, 61102A000101BFX, and 62787A00101EFX. This work was performed while DLD held a National Research Council-Naval Medical Research Institute Research Associateship.

References

Blackman MJ, Heidrich HG, Donachie S, McBride JS, Holder AA (1990) A single fragment of a malaria merozoite surface protein remains on the parasite during red cell invasion and is the target of invasion-inhibiting antibodies. J Exp Med 172:379–382

Burns JM, Majarian WR, Young JF, Daly TM, Long CA (1989) A protective monoclonal antibody recognizes an epitope in the carboxyl-terminal cysteine-rich domain in the precursor of the major merozoite surface antigen of the rodent malarial parasite, *Plasmodium yoelii*. J Immunol Methods 143:2670–2676

Chapman BS, Thayer RM, Vincent KA, Haigwood NL (1991) Effect of intron A from human cytomeglavirus (Towne) immediate-early gene on heterologous expression in mammalian cells. Nucleic Acids Research 14:3979–3986

Charoenvit Y, Mellouk S, Cole C, Bechara R, Leef MF, Sedegah M, Yuan LF, Robey FA, Beaudoin RL, Hoffman SL (1991) Monoclonal, but not polyclonal, antibodies protect against *Plasmodium yoelii* sporozoites. J Immunol 146:1020–1025

Charoenvit Y, Mellouk S, Sedegah M, Toyoshima T, Leef MF, de la Vega P, Beaudoin RL, Aikawa M, Fallarme V, Hoffman SL (1995) 17 kDa *Plasmodium yoelii* hepatic and erythrocytic stage protein is the target of protective antibodies. Exp Parasitol 80:419–429

Collins, WE (1994) In: Baer JF, Weller RE, Kakoma I (eds) *Aotus*: The owl monkey. Academic, New York. pp 245–258

Cullen BR (1986) Trans-activation of human immunodeficiency virus occurs via a bimodal mechanism. Cell 46:973–982

Daly TM, Long CA (1993) A recombinant 15-kilodalton carboxyl-terminal fragment of *Plasmodium yoelii yoelli* 17XL merozoite surface protein 1 induces a protective immune response in mice. Infect Immun 61:2462–2467

Daly TM, Long CA (1995) Humoral response to a carboxyl-terminal region of the merozoite surface protein-1 plays a predominant role in controlling blood-stage infection in rodent malaria. J Immunol 155:236–243

Danko I, Fritz JD, Jiao S, Hogan K, Latendresse JS, Wolff JA (1994) Pharmacological enhancement of in vivo foreign gene expression in muscle. Gene Therapy 1:114–121

Diggs CL, Ballou WR, Miller LH (1993) The major merozoite surface protein as a malaria vaccine target. Parasitol Today 9:300–302

Doolan DL, Sedegah M, Hedstrom RC, Hobart P, Charoenvit Y, Hoffman SL (1996a) Circumventing genetic restriction of protection against malaria with multi-gene DNA immunization: $CD8^+$ T cell, interferon-γ, nitric oxide dependent immunity. J Exp Med 183:1739–1746

Doolan DL, Hedstrom RC, Rogers WO, Charoenvit Y, Rogers M, de la Vega P, Hoffman SL (1996b) Characterization of the protective hepatocyte erythrocyte protein 17 kDa gene of *Plasmodium yoelii*: homolog of *P. falciparum* exported protein-1. J Biol Chem 271:17861–17868

Doolan DL, Hoffman SL (1997) Multi-gene vaccination against malaria: a multi-stage, multi-immune response approach. Parasitol Today 13:171–178

Etlinger HM, Caspers P, Matile H, Schoenfeld HJ, Stueber D, Takacs B (1991) Ability of recombinant or native proteins to protect monkeys against heterologous challenge with *Plasmodium falciparium*. Infect Immun 59:3498–3503

Farley PJ, Srivastava R, Long C (1994) Sequence of the gene encoding the N-terminal portion of the *Plasmodium yoelii yoelii* 17XL merozoite surface protein-1 (MSP-1). Gene 151:335–336

Flynn JL, Weiss WR, Norris KA, Seifert HS, Kumar S, So M (1990) Generation of a cytotoxic T-lymphocyte response using a *Salmonella* antigen-delivery system. Mol Microbiol 4:2111–2118

Freeman RR, Holder AA (1983) Characteristics of the protective response of BALB/c mice immunized with a purified *Plasmodium yoelii* schizont antigen. Clin Exp Immunol 54:609–616

Good MF (1994) Antigenic diversity and MHC genetics in sporozoite immunity. Immunol Lett 41:95–98

Gramzinski RA, Maris DM, Obaldia N, Rossan R, Sedegah M, Wang R, Hobart P, Margalith M, Hoffman SL (1996) Optimization of antibody responses to a *Plasmodium falciparum* DNA vaccine in *Aotus* monkeys. Vaccine Res 5:173–183

Hartikka J, Sawdey M, Cornefert-Jensen F, Margalith M, Barnhart K, Vahlsing L, Meek J, Marquet M, Hobart P, Norman J, Manthorpe M (1996) An improved plasmid DNA expression vector for direct injection into skeletal muscle. Human Gene Ther 7:1205–1217

Hedstrom RC, Sedegah M, Hoffman SL (1994) Prospects and strategies for development of DNA vaccines against malaria. Res Immunol 145:476–483

Hoffman SL, Isenbarger D, Long GW, Sedegah M, Szarfman A, Waters L, Hollingdale MR, van der Miede PH, Finbloom DS, Ballou WR (1989) Sporozoite vaccine induces genetically restricted T cell elimination of malaria from hepatocytes. Science 244:1078–1081

Hoffman SL, Weiss W, Mellouk S, Sedegah M (1990) Irradiated sporozoite vaccine induces cytotoxic T lymphocytes that recognize malaria antigens on the surface of infected hepatocytes. Immunol Letters 25:33–38

Hoffman SL, Franke ED, Hollingdale MR, Druilhe P (1996) Attacking the infected hepatocyte. In: Hoffman SL (ed) Malaria vaccine development. ASM, Washington DC, pp 35–76

Hoffman SL, Crutcher JM, Puri SK, Ansari AA, Villinger F, Franke ED, Singh PP, Finkelman F, Gately MK, Dutta GP, Sedegah M (1997) Sterile protection of monkeys against malaria after administration of interleukin-12. Nature Med 3:80–83

Holder AA, Freeman RR (1981) Immunization against blood-stage rodent malaria using purified parasite antigens. Nature 294:361–364

Khusmith S, Charoenvit Y, Kumar S, Sedegah M, Beaudoin RL, Hoffman SL (1991) Protection against malaria by vaccination with sporozoite surface protein 2 plus CS protein. Science 252:715–718

Khusmith S, Sedegah M, Hoffman SL (1994) Complete protection against *Plasmodium yoelii* by adoptive transfer of a $CD8^+$ cytotoxic T cell clone recognizing sporozoite surface protein 2. Infect Immun 62:2979–2983

Lal AA, De La Cruz VF, Welsh JA, Charoenvit Y, Maloy WL, McCutchan TF (1987) Structure of the gene encoding the circumsporozoite protein of *Plasmodium yoelii*. J Biol Chem 262:2937–2940

Lee KP, June CH, Thompson CB (1991) The CD28 signal transduction pathway in T cell activation. Advances in regulation of cell growth, 2:141–160

Lewis AP (1989) Cloning and analysis of the gene encoding the 230-kilodalton merozoite surface antigen of *Plasmodium yoelli*. Mol Biochem Parasitol 36:271–282

Li S, Rodrigues M, Rodriguez D, Rodriguez JR, Esteban M, Palese P, Nussenzweig RS, Zavala F (1993) Priming with recombinant influenza virus followed by administration of recombinant vacinia virus induces $CD8^+$ T-cell-mediated protective immunity against malaria. Proc Natl Acad Sci USA 90:5214–5218

Ling IT, Ogun SA, Holder AA (1994) Immunization against malaria with a recombinant protein. Parasite Immunol 16:63–67

Luke CJ, Carner K, Liang X, Barbour AG (1997) An OspA-based DNA vaccine protects mice against infection with Borrelia burgdorferi. J Infect Dis 175:91–97

Manthorpe M, Cornefert Jensen F, Hartikka J, Felgner J, Rundell A, Margalith M, Dwarki V (1993) Gene therapy by intramuscular injection of plasmid DNA: studies on firefly luciferase gene expression in mice. Human Gene Therapy 4:419–431

Panina Bordignon P, Tan A, Termijtelen A, Demotz S, Corradin G, Lanzavecchia A (1989) Universally immunogenic T cell epitopes: promiscuous binding to human MHC class II and promiscuous recognition by T cells. Eur J Immunol 19:2237–2242

Pearce E, Caspar P, Grzych J, Lewis FA, Sher A (1991) Downregulation of Th1 cytokine production accompanies induction of Th2 responses by a parasitic helminth, *Schistosoma mansoni*. J Exp Med 173:159–166

Perales JC, Ferkol T, Beegen H, Ratnoff OD, Hanson RW (1994) Gene transfer In vivo: sustained expression and regulation of genes introduced into the liver by receptor-targeted uptake. Proc Natl Acad Sci USA 91:4086–4090

Perrin LH, Merkli B, Loche M, Chizzolini C, Smart J, Richle R (1984) Antimalarial immunity in Saimiri monkeys. Immunization with surface components of asexual blood stages. J Exp Med 160:441–451

Raz E, Watanabe A, Baird SM, Eisenberg RA, Parr TB, Lotz M, Kipps TJ, Carson DA (1993) Systemic immunological effects of cytokine genes injected into skeletal muscle. Proc Natl Acad Sci USA 90:4523–4527

Renia L, Grillot D, Marussig M, Corradin G, Miltgen F, Lambert PH, Mazier D, Del Giudice G (1993) Effector functions of circumsporozoite peptide-primed $CD4^+$ T cell clones against *Plasmodium yoelii* liver stages. J Immunol 150:1471–1478

Rodrigues MM, Cordey AS, Arreaza G, Corradin G, Romero P, Maryanski JL, Nussenzweig RS, Zavala F (1991) $CD8^+$ cytolytic T cell clones derived against the *Plasmodium yoelii* circumsporozoite protein protect against malaria. Int Immunol 3:579–585

Rodrigues MM, Li S, Murata K, Rodriguez D, Rodriguez JR, Bacik I, Bennink JR, Yewdell JW, Garcia Sastre A, Nussenzweig RS, Esteban M, Palese P, Zavala F (1994) Influenza and vaccinia viruses expressing malaria $CD8^+$ T and B cell epitopes. Comparison of their immunogenicity and capacity to induce protective immunity. J Immunol 153:4636–4648

Rogers WO, Rogers MD, Hedstrom RC, Hoffman SL (1992) Characterization of the gene encoding sporozoite surface protein 2, a protective Plasmodium yoelii sporozoite antigen. Mol Biochem Parasitol 53:45–52

Romero P, Maryanski JL, Corradin G, Nussenzweig RS, Nussenzweig V, Zavala F (1989) Cloned cytotoxic T cells recognize an epitope in the circumsporozoite protein and protect against malaria. Nature 341:323–325

Schofield L, Villaquiran J, Ferreira A, Schellekens H, Nussenzweig RS, Nussenzweig V (1987) Gamma-interferon, $CD8^+$ T cells and antibodies required for immunity to malaria sporozoites. Nature 330:664–666

Sedegah M, Beaudoin RL, de la Vega P, Leef MF, Ozcel MA, Jones E, Charoenvit Y, Yuan LF, Gross M, Majarian WR, Robey FA, Weiss WR, Hoffman SL (1988) Use of a vaccinia construct expressing the circumsporozoite protein in the analysis of protective immunity to Plasmodium yoelii. In: L Lasky (ed) Technological advances in vaccine development. Liss, New York, pp 295–309

Sedegah M, Beaudoin RL, Majarian WR, Cochran MD, Chiang CH, Sadof J, Aggarwal A, Charoenvit Y, Hoffman SL (1990) Evaluation of vaccines designed to induce protective cellular immunity against the I circumsporozoite protein: vaccinia, pseudorabies, and salmonella transformed with circumsporozoite gene. Bull World Health Organ 68 (Suppl): 109–114

Sedegah M, Chiang CH, Weiss WR, Mellouk S, Cochran MD, Houghton RA, Beaudoin RL, Smith D, Hoffman SL (1992) Recombinant pseudorabies virus carrying a plasmodium gene: herpesvirus as a new live viral vector for inducing T- and B-cell immunity. Vaccine 10:578–584

Sedegah M, Finkelman F, Hoffman SL (1994a) Interleukin-12 induction of interferon-gamma-dependent protection against malaria. Proc Natl Acad Sci USA 91:10700–10702

Sedegah M, Hedstrom RC, Hobart P, Hoffman SL (1994b) Protection against malaria by immunization with plasmid DNA encoding circumsporozoite protein. Proc Natl Acad Sci USA 91:9866–9870

Siddiqui WA, Tam LQ, Kramer KJ, Hui GS, Case SE, Yamaga KM, Chang SP, Chan EB, Kan SC (1987) Merozoite surface coat precursor protein completely protects *Aotus* monkeys against *Plasmodium falciparum* malaria. Proc Natl Acad Sci USA 84:3014–3018

Stoute JA, Slaoui M, Heppner DG, Monin P, Kester K, Desmons P, Wellde BT, Garcon N, Krzych U, Marchand M, Ballou WR, Cohen JD (1997) A preliminary evaluation of a recombinant circumsporozoite protein vacine against *Plasmodium falciparum* malaria. N Eng J Med 336:86–91

Valmori D, Pessi A, Bianchi E, Corradin G (1992) Use of human universally antigenic tetanus toxin T cell epitopes as carriers for human vaccination. J Immunol Methods 149:717–721

Wang B, Ugen KE, Srikantan V, Agadjanyan MG, Dang K, Refaeli Y, Sato AI, Boyer J, Williams WV, Weiner DB (1993) Gene inoculation generates immune responses against human immunodeficiency virus type 1. Proc Natl Acad Sci USA 90:4156–4160

Wang R, Charoenvit Y, Corradin G, De la Vega P, Franke ED, Hoffman SL (1996) Protection against malaria by *Plasmodium yoelii* sporozoite surface protein 2 linear peptide induction of CD4+ T cell and IFN-γ dependent elimination of infected hepatocytes. J Immunol 457:4061–4067

Weiss WR, Sedegah M, Beaudoin RL, Miller LH, Good MF (1988) $CD8^+$ T cells (cytotoxic/suppressors) are required for protection in mice immunized with malaria sporozoites. Proc Natl Acad Sc USA 85:573–576

Weiss WR, Good MF, Hollingdale MR, Miller LH, Berzofsky JA (1989) Genetic control of immunity to *Plasmodium yoelii* sporozoites. J Immunol 143:4263–4266

Weiss WR, Mellouk S, Houghten RA, Sedegah M, Kumar S, Good MF, Berzofsky JA, Miller LH, Hoffman SL (1990) Cytotoxic T cells recognize a peptide from the circumsporozoite protein on malaria-infected hepatocytes. J Exp Med 171:763–773

Weiss WR, Berzofsky JA, Houghten RA, Sedegah M, Hollingdale MR, Hoffman SL (1992) A T cell clone directed at the circumsporozoite protein which protects mice against both *P. yoelii* and *P. berghei*. J Immunol 149:2103–2109

WHO (1990) Tropical diseases in media spotlight. W.H.O. TDR News 31:3

Xiang Z, Ertl HCJ (1995) Manipulation of the immune response to a plasmid-encoded viral antigen by coinoculation with plasmids expressing cytokines. Immunity 2:129–135

DNA-Based Immunization Against Hepatitis B: Experience with Animal Models

H.L. DAVIS

1 Hepatitis B Virus: A Serious Health Problem

1.1 Hepatitis B Virus Infection

Infection with the hepatitis B virus (HBV) continues to present a worldwide health problem. Whether the escalating infection and disease will eventually be controlled depends largely on the widespread availability of safe, effective and affordable vaccines (PURCELL 1994).

A large proportion of individuals with acute infection (80%) have a subclinical case and as such are often unaware that they are infected, whereas medical management may be required for the 20% who develop acute hepatitis. Only a very few individuals (< 1%) present with a fulminant hepatitis, which can be fatal unless a liver transplant is performed. Regardless of the clinical course a certain proportion of infected individuals will fail to launch the appropriate immune responses to clear

Loeb Research Institute, Ottawa Civic Hospital, 725 Parkdale Avenue, Ottawa, Canada, K1Y 4E9, and Faculties of Health Sciences and Medicine, University of Ottawa the virus and as such they become chronic HBV carriers. In areas of the world where HBV is endemic, such as sub-Saharan Africa and much of Asia, a significant proportion of the population (15%–20%) is chronically infected with HBV. In these areas transmission is usually vertical, from a chronically infected mother to her newborn child, and as many as 90–95% of these cases of perinatal infection will progress to the chronic carrier state. In contrast, in developed areas of the world transmission is usually horizontal between adults, principally by transfer of blood or blood products (e.g., needle-stick injuries) or other body fluids (e.g., sexual contact). The majority of individuals infected as adults will recover completely and are thereafter protected against subsequent infection; however about 5%–10% do become chronic carriers (Ellis 1993; Tiollais and Buendia 1991). It is estimated that there are over 300 million chronic HBV carriers in the world today and although some of them are healthy and without liver pathology, approximately 70% remain in a state of chronic hepatitis which renders them particularly susceptible to developing cirrhosis and hepatocellular carcinoma later in life (Buendia 1992). There is as yet no cure for chronic HBV infection.

1.2 Immunity to Hepatitis B Virus Surface Antigen

Although both humoral and cell-mediated immunity result from natural HBV infection, the presence of antibodies alone is sufficient to confer protection from infection or reinfection. The exact role of cytotoxic T lymphocytes (CTLs) in protection against or clearance of infection is not certain.

The HBV genome contains only four genes: S, C, X and P, two of which encode for structural proteins (Ganem and Varmus 1987). The C gene codes for the nucleocapsid or core protein (HBc) and the S gene encodes the envelope protein. The S gene is a single, large open reading frame with three in-frame ATG start codons such that it is divided into three distinct regions. These are called the preS1, preS2 and S domains (proceeding in a 5′ to 3′ direction). Each of these domains contains B cell epitopes which are known collectively as HBV surface antigen (HBs). The different sized polypeptides produced are known as small or major (S), middle (M = pre-S2 + S) and large (L = preS1 + preS2 + S) envelope proteins (Fig. 1). Furthermore, since each can exist in glycosylated or unglycosylated forms there are a total of six different HBs-containing polypeptides. The envelope of the infectious 42 nm HBV particle (called the Dane particle) contains all six forms, with a predominance of the S forms and variable amounts of the M and L forms. The serum of infected individuals also contains large numbers of smaller (22 nm) round or filamentous empty subviral particles composed solely or predominantly of S and sometimes small amounts of M protein (Tiollais and Buendia 1991).

Both anti-HBs and anti-HBc antibodies are detected after natural infection. There is a very clear role for anti-HBs in conferring protective immunity and all vaccines used in humans to date have been designed to elicit this. The common clinical standard for anti-HBs antibody levels is measured as milli-international units/ml (mIU/ml) and in humans a level of 10 mIU/ml is considered sufficient to confer

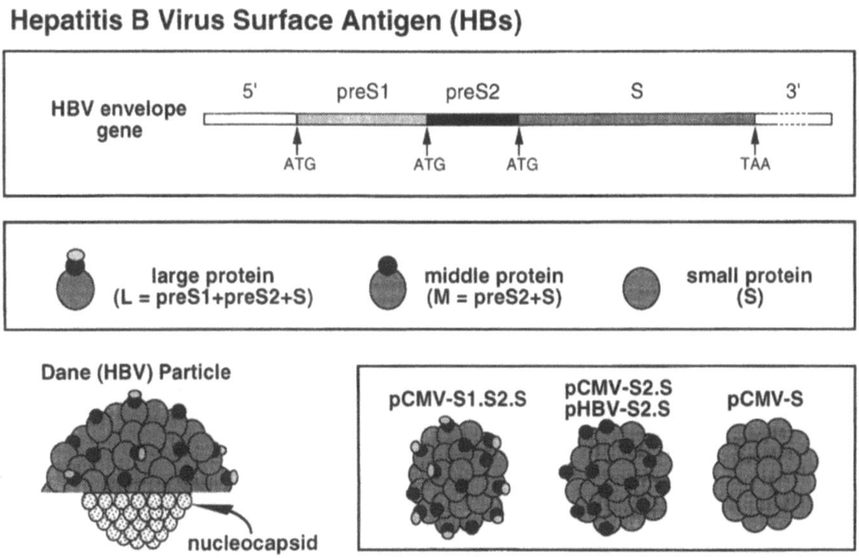

Fig. 1. The hepatitis B virus (HBV) envelope protein is encoded by a single gene which is divided into S, pre-S2 and pre-S1 regions by three in-frame ATG start codons. Six different polypeptides are produced: glycosylated and unglycosylated forms of the small (S), middle (M = pre-S2 + S) and large (L = pre-S1 + pre-S2 + S) envelope proteins. The envelope of the infectious HBV (Dane particle) is composed predominantly of S with some M and L forms. We have constructed DNA expression vectors containing different parts of the envelope protein gene which can produce one or more forms of envelope protein since internal initiation can occur at any of the included ATG codons. Three constructs use the cytomegaloviral (CMV) promoter to drive expression of the S gene (pCMV-S which expresses S protein), the pre-S2 + S gene (pCMV-S2.S expressing M + S protein) or the pre-S1 + pre-S2 + S gene (pCMV-S1.S2.S expressing L + M + S protein). A fourth construct contains the entire envelope gene but expresses only the pre-S2 + S regions (M + S proteins) as it is driven by an endogenous HBV promoter situated within the pre-S1 domain (pHBV-S2.S)

protection (Centers for Disease Control 1987). The role for anti-HBc in conferring protective immunity is less clear. HBc is highly immunogenic but the protective effect of the anti-HBc antibodies is questioned since chronic carriers may have high titers of anti-HBc but not anti-HBs. Furthermore, high levels of maternal anti-HBc in infants born to such chronic carrier mothers fail to protect them from infection (Purcell and Gerin 1987). By contrast, immunization to HBc alone has been shown to protect chimpanzees against challenge with live HBV, but it is possible that HBc-specific CTLs play a role in this situation (Iwarson et al. 1985; Murray et al. 1984).

2 Vaccines for HBV: Past and Present

2.1 Plasma-Derived Vaccines

The first HBV vaccine to be applied to humans involved injection of empty 22 nm subviral particles which had been purified from the plasma of chronic carriers.

Several such vaccines have been developed which differ primarily in the methods used for purifying the particles and inactivating the plasma to kill any infectious contaminants (MAUPAS et al. 1976; PURCELL and GERIN 1975). Although these plasma-derived subunit vaccines are safe and efficacious and are still used in some countries, they are not suitable for widespread use owing to the limited supply of chronically infected human plasma, the high cost of production and the need to test each batch on chimpanzees for safety. In addition, there are persistent fears that the plasma-derived vaccines pose a risk of transmitting HBV itself or other infectious diseases such as HIV, and although such concerns are unfounded they do reduce the acceptability of the plasma-derived vaccines (MAUGH 1980; PURCELL and GERIN 1987; STEPHENNE 1988).

2.2 Recombinant Antigen Vaccines

The HBV vaccine currently used in many areas of the world including most European and North American countries consists of recombinant HBV envelope protein. These vaccines contain subviral particles composed either solely of S protein produced by yeast cells (VALENZUELA et al. 1982) or the M plus S proteins produced by Chinese hamster ovary cells (MICHEL et al. 1984). Although they are highly effective, the recombinant protein vaccines are prohibitively expensive for use in many of the areas of the world where HBV is endemic. For example, the cost of vaccinating against HBV can exceed the total cost of vaccinating against all other diseases as recommended by the World Health Organization (WHO). The expense of vaccine delivery is further exacerbated in endemic areas because the HBV vaccine must be initiated within the first days of life to avoid vertical transmission from a chronic carrier mother; however vaccines against the other diseases such as diphtheria, pertussis and tetanus must be given somewhat later to avoid interference by maternal antibodies.

Another problem associated with either the plasma-derived or recombinant subunit vaccines is that of nonresponse (no detectable levels of anti-HBs) or hyporesponse (< 10 mIU/ml anti-HBs) (ELLIS 1993). This situation is particularly worrisome for those individuals who risk exposure to HBV, such as health care workers, persons receiving blood products or partners of chronic HBV carriers. In immunocompetent individuals the overall incidence of nonresponse or hyporesponse is less than 5% in young adults, but increases to 30%–50% with advancing age. The problem of nonresponse is, at least in some cases, HLA-linked and several nonresponsive haplotypes have been identified (KRUSKALL et al. 1992). Other factors which can increase the incidence of nonresponse to HBs include infection with HIV, kidney dialysis and immunosuppression after organ transplant. In these cases the rate of nonresponse can be as high as 80%.

2.3 Other Hepatitis B Virus Vaccines

Various attempts have been made to develop an efficacious and inexpensive vaccine for HBV. One approach has been to express the HBs antigen in a viral vector such as vaccinia (Moss et al. 1984) or adenovirus (LUBECK et al. 1989). Vaccination of chimpanzees with these viral vector vaccines induced significantly lower antibody titers than did the subunit vaccines, and although they protected against hepatitis in most cases, they did not always prevent initial infection. Thus the lower efficacity combined with unpleasant side effects induced by the viral vector itself make it unlikely that these vaccines will replace the plasma-derived or recombinant vaccines, even in developing areas of the world.

Peptide vaccines are another inexpensive approach and have fewer side effects than the viral vectors. They consist of synthetic peptides predicted from the nucleotide sequences of genes of viral structural proteins. Vaccination of chimpanzees with peptides corresponding to defined regions of the S or pre-S domains of the gene for the HBV envelope protein induced partial or complete protection against viral challenge in the majority of animals although antibody titers remained low despite repeated boosts and the use of adjuvants (GERIN et al. 1985; THORNTON et al. 1989). Low immunogenicity is a problem with many peptide vaccines since B cell epitopes are frequently conformational.

Thus there is still a need for an inexpensive, safe and efficacious HBV vaccine for use in developing regions of the world. In addition, there is a need for a vaccine which could overcome the problem of nonresponse to the traditional subunit vaccines, especially for use on individuals at risk of becoming infected. Genetic or DNA-based immunization may meet these needs. Firstly, DNA vaccines should be less expensive to produce than the plasma-derived and recombinant antigen-based vaccines and the cost of delivery to developing regions of the world will be less owing to its heat-stability, which precludes the need for a cold-chain. In addition, costs will be further decreased if fewer boosts are required owing to the longevity of *in vivo* gene expression.

3 DNA-Based Immunization to Hepatitis B Virus in Mice

We have developed a model of DNA-based immunization against HBV by intramuscular injection of HBs-expressing recombinant plasmid DNA (DAVIS et al. 1993). Three of the eukaryotic expression vectors use the immediate early promoter of cytomegalovirus (CMV) to drive expression of the S (pCMV-S), the preS2 + S (pCMV-S2.S) or the preS1 + preS2 + S (pCMV-S1.S2.S) genes of the HBV envelope. The HBs-containing proteins expressed from these vectors are S, S + M and S + M + L respectively. A fourth vector uses the endogenous HBV promoter within the pre-S1 domain to drive expression of the S + M proteins (pHBV-S2.S) (21) (Fig. 1).

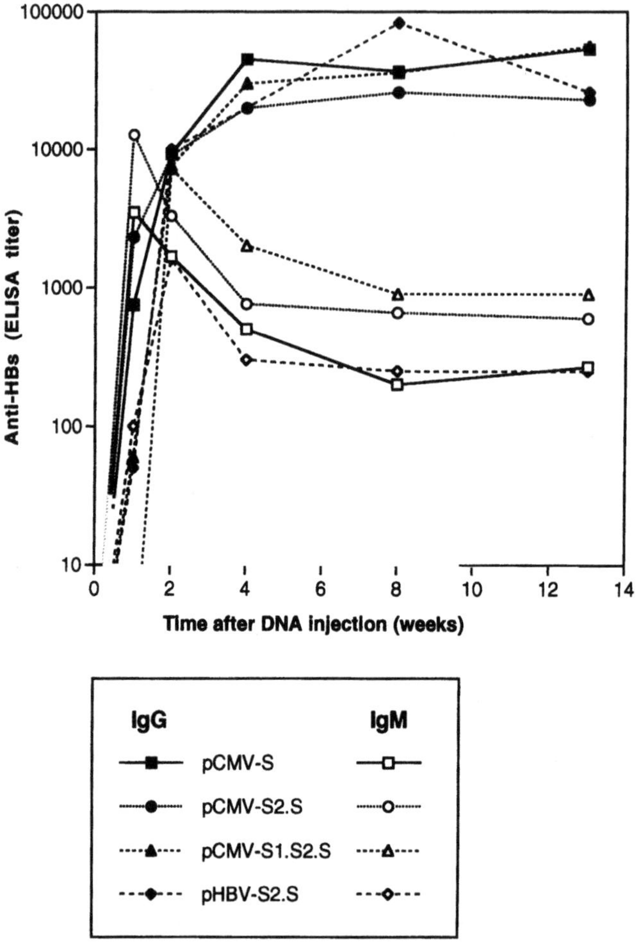

Fig. 2. Kinetics of total IgM and IgG anti-HBs antibodies in mice immunized with different hepatitis B virus (HBV) envelope-expressing plasmids. Antibodies were detected in pooled sera from mice injected with the same construct ($n = 8$) taken at the same time point after a single injection of DNA (100 µg) into the tibialis anterior muscle. For the ELISA assay, anti-HBs antibodies were bound to a solid phase of HBs particles of a homologous subtype (ay) containing the pre-S2 domain, and were then detected by addition of peroxidase-labeled goat anti-mouse IgM or anti-mouse IgG. End-point titers were defined as the highest serum dilution that resulted in an absorbance value two times greater than that of nonimmune serum with a cutoff value of 0.05

Intramuscular injection in mice of any of these four plasmid DNA vectors results in strong and sustained humoral and cell-mediated immune responses.

3.1 *In Vivo* Synthesis of Hepatitis B Virus Surface Antigen

Intramuscular injection of 50 µg of HBs-expressing DNA into the tibialis anterior muscle (TA) of the mouse results in transfection of approximately 1%–2% of the muscle fibers, which can be detected by immunofluorescent label. The transfected fibers, which have a normal histological appearance 5 days after DNA injection, are seen to fragment at about 10 days and to have disappeared completely by 30 days. This disappearance of HBs-expressing muscle fibers is likely due to attack by HBs-specific CTLs since similarly transfected fibers are spared in mice with severe combined immunodeficiency (Davis et al., 1997).

Low levels of circulating antigen (≤1 ng/ml) are detected only at early time points (2–4 weeks after DNA injection). The lack of detectable HBs at later times might reflect neutralization of antigen by antibodies (see below) as well as reduced synthesis with the destruction of expressing fibers.

3.2 Humoral Response

Antibodies to HBs may be detected as early as 1 week after injection of DNA. The subsequent humoral response is comparable to that seen following natural infection with respect to kinetics, antibody isotype and fine-specificity of the response. For example, antibodies detected at the earliest time point are principally of the IgM isotype but they soon undergo a class shift to the IgG isotype, which thereafter remains the predominant form. Antibodies specific to group and subtype determinants of HBs may be detected, and with the pre-S2-expressing vector there are also antibodies to the pre-S2 domain. No antibodies to the pre-S1 domain were detected with the pCMV-S1.S2.S vector despite the existence of an identified B cell epitope on this region of the L polypeptide. This may be related to poor secretion of L-containing particles (Michel et al. 1995).

3.2.1 Dose-Response

The precocity and strength of the humoral response is dose-dependent. With a single injection of 1 µg DNA into normal muscle, only a few mice seroconvert (defined arbitrarily as >1 mIU/ml) by 8 weeks and these attain only very low titers of anti-HBs. However, in mice injected with 10 or 100 µg HBs-expressing DNA there is 100% seroconversion by 2 and 4 weeks respectively. Antibody titers are higher with greater amounts of DNA but not in a directly proportional manner since those induced with 100 µg DNA are only about 2.5 times higher than those with 10 µg DNA (Davis et al. 1995a). This is likely due to superfluous DNA at the injection site and higher antibody titers might be attained if the DNA was divided and delivered to several different sites.

Transfection efficiency is improved if the DNA is injected into regenerating muscle, which can be obtained by first inducing degeneration by injection of necrotizing agents such as local anaesthetics or snake-venom-derived cardiotoxin. In this case about ten times more muscle fibers are transfected for a given amount of DNA (Davis et al. 1993a, c). Following injection of 100 µg HBs-expressing plasmid DNA into regenerating mouse TA muscle, antibodies are first detected 1 week earlier and reach approximately ten-fold higher titers than with injection into normal TA muscle (Davis et al. 1994). The mean peak antibody titers reached are 10^3 for IgM and 10^5 for IgG by ELISA end-point dilution, although individual animals can attain significantly higher titers. These anti-HBs titers are equivalent to about 10^4 mIU/ml, which is 1000 times greater than the critical 10 mIU/ml level which is considered necessary to confer protection in humans.

3.2.2 Longevity of Response

Peak antibody titers are reached by 4–8 weeks after injection of DNA and are maintained at near maximal levels for up to 17 months without boost. Despite the high sustained response, antibody levels may be further increased approximately ten-fold with a second injection of DNA or about five-fold with administration of recombinant HBs protein (Davis et al. 1996).

3.2.3 Effect of Vector

The kinetics and specificity of the humoral response are influenced by the vector used. Addition of the pre-S2 region (pCMV-S2.S) results in peak titers of IgM and IgG 1 week earlier than with S alone, and the IgM response is considerably stronger (titer approximately 10^4). B and T helper cell epitopes have previously been identified on the pre-S2 polypeptide. In contrast, the DNA vector expressing middle protein from the endogenous promoter (pHBV-S2.S) produced a much slower humoral response, with significant levels of IgM and IgG not being detected until 4 weeks after DNA injection. Nevertheless, this construct ultimately resulted in the highest antibody titers (titer approximately 10^6). Addition of the pre-S1 region also results in a slower immune response than with S alone, since significant levels of IgM are not detected until 2 weeks after DNA injection. This delay may be due to the known poor secretion of the L protein.

3.3 Cytotoxic T Lymphocyte Response

DNA-based immunization by injection of HBs-expressing plasmid DNA also results in a very strong CTL response. Mice immunized with vectors expressing all three forms of HBs protein gave rise to high levels of both CTLs and their precursors. For example, after specific restimulation *in vitro* for 5 days with HBs-expressing cells, spleen cells from DNA immunized mice were capable of specific lysis of 80%–90% at effector:target ratios as low as 2.5:1. Even after nonspecific *in vitro* stimulation (overnight with interleukin-2), the spleen cells exhibited 40% specific lysis at effector:target ratios of 200:1. CTL activity, which was first detected 6 days after injection of DNA, reached maximal levels by 12 days and was maintained for at least several months (Davis et al. 1995b).

3.4 Haplotype-Restricted Nonresponse to Hepatitis B Virus Surface Antigen

Certain strains of mice are known to be poor responders to the B cell epitopes of the S protein unless polypeptides encoded by the pre-S2 and/or pre-S1 domains of the envelope gene are also present (MILICH 1988). This hyporesponsiveness to S protein, which is genetically linked to the H-2 haplotype, can be overcome by DNA-based immunization. When congenic strains of mice of different H-2 haplotypes were immunized with either S-expressing DNA or with recombinant S protein with adjuvant, all mice responded after a single injection of DNA whereas with antigen-based immunization the poor responding strains of mice required a protein boost before antibodies were detected (Brazolot-Millan, Michel, Mancini, Whalen and Davis, unpublished results). These findings are of potential clinical significance for nonresponse or hyporesponse in humans.

3.5 DNA-Based Immunization of Transgenic Mice

Transgenic mice expressing HBs can be used as a model of the HBV chronic carrier. One such transgenic mouse line has the HBV genome minus the core gene as a transgene which is expressed almost exclusively in the liver. These mice produce high levels of circulating subviral particles containing S and M HBV envelope proteins, yet do not produce any anti-HBs antibodies (BABINET et al. 1985). DNA-based immunization by intramuscular injection with the pCMV-S2.S vector results in a rapid appearance of anti-HBs antibodies and concomitant decrease in circulating antigen. The loss of circulating antigen is certainly due in part to neutralization by antibodies but is also due to down-regulation of transgene expression in the liver, a phenomenon which seems to be mediated by a noncytolytic mechanism (Mancini et al., 1996). Injection of pure recombinant HBs protein has no effect on these mice unless it contains a heterologous epitope (e.g., HIV V3 loop) as a fusion protein (MANCINI et al. 1993). It is interesting that DNA-based immunization can break tolerance or anergy to HBs in these transgenic mice when it is considered that the DNA encodes the identical gene product as the transgene and that the amount of antigen expressed in the muscle from the injected DNA would be a very small fraction of that synthesized in the liver from the transgene. It is likely due to the "danger" signal provided by immunostimulatory Cp6 motifs in the plasmid DNA, which are known to promote antigen presentation in conjunction with costimulatory molecules (Krieg et al., 1995). These results may have important clinical significance for immunotherapeutic treatment of HBV chronic carriers.

4 DNA-Based Immunization to Hepatitis B Virus in Primates

Chimpanzees are an ideal animal model to evaluate DNA vaccines for their potential use in humans since these animals are similar to humans with respect to susceptibility to HBV infection and antibody titers required for protection. Furthermore, it is the only animal model which can be challenged with human HBV (Prince 1981). We have vaccinated two chimpanzees against hepatitis B virus (HBV) by intramuscular injection of either 2 mg or 400 μg of HBs-expressing plasmid DNA (pCMV-S2.S) at time points 0, 8, 16 and 27 weeks (Davis, McCluskie and Purcell, unpublished results). The chimpanzee injected with 2 mg DNA at each time point attained high levels of antibody (> 100 mIU/ml) after the initial injection of DNA and these were further increased to up to 14000 mIU/ml after the three boosts. The chimpanzee receiving 400 μg DNA had no detectable antibodies until after the second injection of DNA, and although antibody titers of 60 mIU/ml were attained these were transient. Nevertheless, this chimpanzee showed a strong anamnestic response to injection of recombinant HBs given 1 year after the initial DNA injection. As seen with the mice, the antibodies induced by the DNA were against group and subtype determinants of the S protein as well as against the pre-S2 domain. These antibodies were initially of the IgM isotype and subsequently shifted to IgG antibodies, which were predominantly IgG1. The antibody titers induced by the DNA vaccine in the strongly responding chimpanzee were almost as high as the best levels detected among six chimpanzees receiving a series of three injections of commercial recombinant subunit vaccine (Davis et al., 1996).

These results indicate that the DNA approach is feasible for prophylactic or therapeutic immunization of humans against HBV. Nevertheless, it would be preferable that the approach be optimized such that lower doses of DNA could be used. Optimization could involve increasing the efficiency of gene transfer, improving the level of gene expression or increasing the immune response to the expressed antigen (Davis et al. 1994). This might be accomplished by using different routes of DNA administration (e.g., intramuscular, intradermal, gene-gun) alone or in combination, using facilitators to improve gene transfer, coexpressing the antigen with costimulatory molecules or cytokines and using adjuvants including CpG immunostimulatory motifs.

Acknowledgements. I am grateful to all of my colleagues who have graciously collaborated with me on the studies reported here. I am particularly grateful to Robert G. Whalen (CNRS URA-1115, College de France, Paris, FR) and to Marie-Louise Michel and Maryline Mancini (INSERM U163, Pasteur Institute, Paris, FR), with whom many of these studies have been carried out. I also wish to thank Robert Purcell (NIAID, NIH, Bethesda, MD), Martin Schleef (Qiagen, Hilden, Germany), Reinhold Schirmbeck and Jörg Reimann (University of Ulm, Germany), and Simon Watkins (University of Pittsburgh, PA).

References

Babinet C, Farza H, Morello D, Hadchouel M, Pourcel C (1985) Specific expression of hepatitis B surface antigen (HBsAg) in transgenic mice. Science 230:1160–1163

Buendia MA (1992) Hepatitis B viruses and hepatocellular carcinoma. Adv Cancer Res 59:167–226
Centers for Disease Control (1987) Recommendations of the immunization practises advisory committee. Update on hepatitis B prevention. MMWR 36:353
Davis, HL, Brazolot Millan CL, Watkins SC (1997) Immune-mediated destruction of transfected muscle fibers after direct gene transfer with antigen-expressing plasmid DNA. Gene Ther 4:181–188
Davis HL, Demeneix BA, Quantin B, Coulombe J, Whalen RG (1993) Plasmid DNA is superior to viral vectors for direct gene transfer in adult mouse skeletal muscle. Human Gene Ther 4:733–740
Davis HL, Mancini M, Michel M-L, Whalen RG (1996) DNA-mediated immunization to hepatitis B surface antigen: Longevity of primary response and effect of boost. Vaccine 14:910–915
Davis HL, McCluskie MJ, Gerin JL, Purcell RH (1996) DNA vaccine for hepatitis B: immunogenicity in chimpanzees and comparison with other vaccines. Proc Natl Acad Sci USA 93:7213–7218
Davis HL, Michel M-L, Mancini M, Schleef M, Whalen RG (1994) Direct gene transfer in muscle with plasmid DNA for the purpose of nucleic acid immunization. Vaccine 12:1503–1509
Davis HL, Michel M-L, Whalen RG (1993) DNA based immunization for hepatitis B induces continuous secretion of antigen and high levels of circulating antibody. Hum Molec Genet 2:1847–1851
Davis HL, Michel M-L, Whalen RG(1995) Use of plasmid DNA for direct gene transfer and immunization. NY Acad Sci 772:21–29
Davis HL, Schirmbeck R, Reimann J, Whalen RG (1995) DNA-mediated immunization in mice induces a potent MHC class-I restricted cytotoxic T lymphocyte response to the hepatitis B envelope protein. Hum Gene Ther, 6:1447–1456
Davis HL, Whalen RG, Demeneix BA (1993) Direct gene transfer into skeletal muscle *in vivo*: Factors affecting efficiency of transfer and stability of expression. Hum Gene Ther 4:151–159
Ellis RW (ed) (1993) Hepatitis B vaccines in clinical practise. Marcel Dekker, New York
Ganem D, Varmus HE (1987) The molecular biology of the hepatitis B viruses. Ann Rev Biochem 56:651–693
Gerin JL, Purcell RH, Lerner RA (1985) Use of synthetic peptides to identify protective epitopes of the hepatitis B surface antigen. In: R Lerner, F Brown and R Chanock (eds) Vaccines 85: Modern approaches to vaccines. Molecular and chemical basis of resistance to viral, bacterial and parasitic diseases. Cold Spring Harbour Laboratory, Cold Spring Harbour, New York pp 235–239
Iwarson S, Tabor E, Thomas HC, Goodall A, Waters J, Snoy P, Shih JWK, Gerety RJ (1985) Neutralization of hepatitis B virus infectivity by a murine monoclonal antibody: an experimental study in the chimpanzee. J Med Virol 16:89–96
Krieg AM, Yi AK, Matson S, Waldschmidt TJ, Bishop GA, Teasdale R, Koretzky GA, Klinman DM (1995) CpG motifs in bacterial DNA trigger direct B-cell activation, Nature, 374:546–549
Kruskall MS, Alper CA, Awdeh Z, Yunis EJ, Marcus-Bagley D (1992) The immune response to hepatitis B vaccine in humans: inheritance patterns in families. J Exp Med 175:495–502
Lubeck MD, Davis AR, Chengalvala M, Natuk RJ, Morin JE, Molnar-Kimber K, Mason BB, Bhat B M, Mizutani S, Hung PP, Purcell RH (1989) Immunogenicity and efficacy testing in chimpanzees of an oral hepatitis B vaccine based on live recombinant adenovirus. Proc Natl Acad Sci USA 86:6763–6767
Mancini M, Hadchouel M, Davis HL, Whalen RG, Tiollais P, Michel ML (1996) DNA-mediated immunization breaks tolerance in a transgenic mouse model of hepatitis B surface antigen chronic carriers. Proc Natl Acad Sci USA 93:12496–12501
Mancini M, Hadchouel M, Tiollais P, Pourcel C, Michel ML (1993) Induction of anti-hepatitis B surface antigen (HBsAg) antibodies in HBsAg producing transgenic mice: a possible way of circumventing "nonresponse" to HBsAg. J Med Virol 39:67–74
Maugh TH (1980) Hepatitis B vaccine passes first major test. Science 210:760–762
Maupas P, Goudeau A, Coursaget P, Drucker J (1976) Immunization against hepatitis B in man. Lancet 1:1367–1370
Michel ML, Pontisso P, Sobczak E, Malpièce Y, Streeck RE, Tiollais P (1984) Synthesis in animal cells of hepatitis B surface antigen particles carrying a receptor for polymerized human serum albumin. Proc Natl Acad Sci USA 81:7708–7712
Michel ML, Davis HL, Schleef M, Mancini M, Tiollais P, Whalen RG (1995) DNA-mediated immunization to the hepatitis B surface antigen in mice: aspects of the humoral response mimic hepatitis B viral infection in humans. Proc Natl Acad Sci USA 92:5307–5311
Milich DR (1988) T- and B-cell recognition of hepatitis B viral antigens. Immunol Today 9:380–386
Moss B, Smith GL, Gerin JL, Purcell RH (1984) Live recombinant vaccinia virus protects chimpanzees against hepatitis B. Nature 311:67–69
Murray K, Bruce SA, Hinnen A, et al. (1984) Hepatitis B virus antigens made in microbial cells immunize against viral infection. Embo J 3:645–650

Prince AM (1981) The use of chimpanzees in biomedical research. In: US Department of Health and Human Services (ed) Trends in bioassay methodology in vivo, in vitro and mathematical approaches. vol 82, NIH, Bethesda, pp 81–98

Purcell RH (1994) Hepatitis viruses: changing patterns of human disease. Proc Natl Acad Sci USA 91: 2401–2406

Purcell RH, Gerin JL (1975) Hepatitis B subunit vaccine: a preliminary report of safety and efficacy tests in chimpanzees. Am J Med Sci 270:395–399

Purcell RH, Gerin JL (1987) Hepatitis B vaccines, past, present and future: lessons for AIDS vaccine development. In: Robinson W, Koike K, Will H (eds) Hepadna viruses, Alan R Liss, New York, pp 465–479

Stephenne J (1988) Recombinant versus plasma-derived hepatitis B vaccines: issues of safety immunogenicity and cost-effectiveness. Vaccine 6:299–303

Thornton GB, Moriarty AM, Milich DR, Eichberg JW, Purcell RH, Gerin JL (1989) Protection of chimpanzees from hepatitis B virus infection after immunization with synthetic peptides: identification of protective epitopes in the pre-S region. In: Chanock RM, Lerner RA, Brown F, Ginsberg H (eds)Vaccines 89, Cold Spring Harbor Laboratory, Cold Spring Harbor, pp 467–471

Tiollais P, Buendia M-A (1991) Hepatitis B virus. Sci Amer 264:48–54

Valenzuela P, Medina A, Rutter WJ, Ammerer G, Hall BD (1982) Synthesis and assembly of hepatitis B virus surface antigen particles in yeast. Nature 298:347–350

DNA Vaccines and Immunity to Herpes Simplex Virus

B.T. Rouse[1], S. Nair[2], R.J.D. Rouse[1], Z. Yu[1], N. Kuklin[1], K. Karem[1], and E. Manickan[1]

1 Background

Infections by herpes simplex virus (HSV) represent an expensive public health problem. Although only rarely a cause of mortality, HSV infections usually cause painful and often distressing lesions and are particularly troublesome since symptomatic recurrent disease is a common outcome once an individual has been infected. Recurrent lesions on the face and genitalia are the most common expression but in some locations, such as the eye, distressing results such as blindness can occur. As regards HSV infections, in 1985 the Committee on Issues and Priorities for New Vaccine Development of the Institute of Medicine set the following goals: a 50% reduction in symptomatic primary infection, a 75% reduction in the number of recurrences and a 60% reduction in the severity of episodes.

Currently in the USA, no licensed vaccines exist for HSV, yet the search for suitable vaccines has gone on unabated for decades. Given the considerable success of vaccines against several animal herpesviruses, one might wonder why the same is not true for HSV. A major reason for this lies with the fact that vaccines against infectious agents work most effectively when used prophylactically. For the control of human HSV infection the greatest need is for vaccines which will diminish the consequences in those who are already exposed to HSV and are carrying the virus in a latent form. Unfortunately, most latently infected individuals suffer periodic reactivation and disease reexpression. Those subject to this situation clamor for therapeutic vaccines which will minimize or even abolish clinical expression.

[1]Department of Microbiology, M409 Walters Life Sciences Building, University of Tennessee, Knoxville, TN 37996-0845, USA
[2]Current address: Duke University Medical Center, Department of Surgery, Durham NC, 27710, USA

Although many candidate vaccines have been touted to fulfill such a role, none have been proven to be satisfactory upon independent double blind evaluation (Stanberry 1996). Moreover, in the HSV field the placebo effect is a well known complication accounting for up to 35% efficacy in some studies (Kutinova et al. 1988). Recently enthusiasm for therapeutic vaccines was rearoused by the results of Straus et al. (1984), demonstrating beneficial effects in a small trial using a glycoprotein vaccine with alum as an adjuvant. This particular trial presented the first controlled data showing that a vaccine can modify the course of a chronic viral infection in human beings. Unfortunately in the study no critical determinant of the immune response which correlated with efficacy was identified. Thus there is a strong suspicion that vaccines might prove effective therapeutically if they optimally stimulate certain components of the immune response. Alas we remain unsure as to the identity of such critical components.

Clearly humans represent inappropriate subjects to evaluate experimental approaches that will define mechanisms of immunity associated with an ideal vaccine. Animal models must be used, although these are all notably imperfect as models of spontaneous recurrent disease. One of the better models to study immunity to HSV is the mouse zosteriform system, and we and others have used this to identify candidate mechanisms responsible for immunity in the skin and nervous system (Simmons and Nash 1984; Mester and Rouse 1991). In the zosteriform model, virus is placed on the scratched flank and spreads via sensory nerves to the dorsal root ganglion where replication occurs. This is followed by dissemination of virus via nerves to the whole dermatome innervated by the dorsal root ganglion. The result is an inflammatory reaction evident clinically 5–7 days later in the whole dermatome. Unfortunately, these lesions rarely heal or at least fail to do so before the mice die of encephalitis, since one consequence of replication in the ganglion is viral spread to the CNS. The zosteriform model is useful to assess immunity and has been used to show immune mechanisms at play. Accordingly it appears that protection against the skin lesions is mediated principally by $CD4^+$ T cells and antibody (Simmons and Nash 1984) whereas spread within the CNS appears to be primarily a function of $CD8^+$ T cells (Simmons and Tscharke 1992). We have extensively used the zosteriform model to evaluate the effectiveness of various vaccine formulations including the use of naked DNA (Manickan et al. 1995a, b, c).

2 General Experimental Design and Immunization Studies

In our studies, all sources of HSV DNA were cloned into the pcDNAI plasmid obtained from Invitrogen (San Diego). This plasmid uses the immediate early promoter of cytomegalovirus (CMV) and in the studies described below was used to drive expression of ICP27 and gB of HSV-1. As controls, vector without insert or the vector driving the LUX gene was used. BALB/c mice were immunized intramuscularly at about 6 weeks of age and again 1 week later with 90 μg DNA and then challenged 7 days after the second dose. We usually chose to use two different challenge doses in the zosteriform model, a low dose consisting of 10 ID_{50} and a high dose of 500 ID_{50}. We used HSV-1 strain 17 for infection. Immunological evaluations were done at day 14. In some experiments, animals were given a single intramuscular injection of 90 µg DNA or were immunized on three occasions intranasally with 100 µg DNA with or without 2 µg cholera toxin. In the intranasal immunization experiments, animals were challenged intravaginally with 3×10^7 pfu of HSV-1 McCrae which represents about 10 ID_{50}. In other experiments, DNA for GM-CSF (kindly supplied by Dr. Hildegund Ertl) expressed in the pRJB-c plasmid and driven by the Rous sarcoma virus (RSV) long terminal repeat (LTR) promoter was used. This DNA was injected into the same site intramuscularly as the DNA that encoded HSV proteins.

Immunization with either ICP27, a regulatory immediate early protein, or gB, a major structural glycoprotein, afforded protection. This was of the order of 70%–80% and was evident against low dose challenge. Neither construct protected against high dose challenge. Control animals immunized with UV-inactivated HSV or with a recombinant vaccinia virus expressing ICP27-2 could resist high dose challenge (Manickan et al. 1995a). Consequently although DNA immunization is effective, as currently used it may prove less efficacious than other approaches. One reason for the apparent diminished efficacy of DNA may be that the response takes longer to develop and reach peak effectiveness than is necessary for other forms of immunization. In support of this idea, we have observed occasional animals challenged after some weeks which do resist high dose infection. In addition, in an experiment in which animals were immunized intramuscularly on a single occasion and then sample animals challenged periodically over a 6 month period, we observed that immunity was not evident until 3 weeks post-immunization (Table 1). Moreover, by 4 weeks, all animals resisted low dose challenge. Sample animals

Table 1. Resistance of mice to herpes simplex virus (HSV) challenge at different times after immunization with gB DNA

post infection Week	Days post challenge	gB DNA	Vector	HSV
1	14	5/5	5/5	0/5
2	14	3/5	5/5	0/5
3	14	1/5	5/5	0/5
4	14	0/5	5/5	0/5
5	14	0/5	5/5	0/5

Mice were immunized on day 0 with 90 µg of gB or vector DNA or 10^7 $TCID_{50}$ HSV-1 (KOS). These animals were zosteriform challenged on 1, 2, 3, 4, 5 weeks post immunization with 10^5 $TCID_{50}$ (100 ID_{50}) of HSV-1.17. Plasmid pcDNAI encoding gB-HSV-1 gene inserted in *Bam*HI and *Hin*d III region under the control of CMV promoter was used as the immunogen. The same plasmid without the inserted gene (negative control) or HSV-1 (KOS) (positive control) were included in the study. Sample of mice challenged after 2, 3, 4, 5, and 6 months of immunization remained resistant to HSV challenge in the gB DNA group. Mice that were vaccinated with gB-DNA were resistant to 10 ID_{50} of HSV-1.17 challenge but those challenged with 50 ID_{50} all developed lesions. High dose virus challenges were performed at 4 weeks p.i. and in another group at month six.

were tested after 4 weeks and although all resisted low dose challenge, all succumbed to challenge with the high infectious dose (Table 1).

We have begun to seek means of enhancing the level of immunity induced following immunization with DNA vaccines. Minimal experiments comparing different routes, such as subcutaneous and intradermal, with intramuscular have led to no improvement. We have not, however, used the popular gene gun which, in the hands of other groups with different systems, has emerged as a superior strategy (FYNAN et al. 1993). Rationalizing that the critical cells involved in antigen presentation and immune induction, are dendritic cells (DCs) two immunization approaches have been used which should theoretically optimally engage such antigen-presenting cells (APCs). In the first approach, enriched populations of splenic DCs isolated from naive mice were transfected in vitro with DNA and after a 3 h induction at 37 °C injected i.m. into mice. A second injection was given after 1 week and sample animals were challenged after a further week. Although by this procedure, certain aspects of immunity were increased, the procedure resulted, with one exception, in resistance to low dose challenge only (Manickan and Rouse, unpublished observations). Interestingly, in animals immunized with DCs that were transfected with both gB and ICP27, some animals (three of five) were resistant to high dose challenge. It remains to be seen if such heightened resistance correlated with any particular immune component.

A second strategy employed is to exploit the fact that the cytokine granulocyte/macrophage colony-stimulating factor (GM-CSF) is known to act as an activator and growth factor for DCs (WITMER-PACK et al. 1987). Accordingly taking the lead from work with rabies virus DNA by XIANG and ERTL (1995), it was expected that the co-injection of viral protein DNA along with GM-CSF DNA might cause enhancement of endogenous DC activity and an elevated level of immunity. As is recorded in Table 2, the approach did indeed succeed in enhancing the activity of certain components of immunity, most notably the $CD8^+$ cytotoxic T lymphocyte (CTL) response and levels of IgG1 antibody. Despite this, few if any animals receiving the gB and GM-CSF DNA combination resisted high dose challenge with HSV using the zosteriform model.

Recently, using a vaginal model to measure immunity to HSV, KRIESEL et al. (1996) immunized with DNA encoding gD of HSV-2 and showed good levels of protection providing vitamin D was used as an adjuvant. In preliminary studies, we have also used mucosal routes of challenge to measure immunity to HSV. Animals that have been immunized intramuscularly with gB DNA do resist challenge to HSV-1 given via the intranasal or intravaginal routes (KUKLIN, 1997). However, to date, challenge has only been measured against a low dose of infection. Intramuscular immunization with gB DNA led to the production of low levels of antibody detectable in vaginal washings, and it is possible that such antibody is responsible for the immunity observed. However, other explanations are also possible with $CD4^+$ T cell responses being perhaps the most likely.

Most interestingly, animals immunized with DNA via the mucosal route were shown to generate immune responses (KUKLIN, 1997). For this purpose animals were lightly anaesthetized and 100 μg DNA was delivered intranasally

Table 2. Enhancement of immunity against herpes simplex virus (HSV) after co-inoculation of gB and GM-CSF DNA

Immunogens given to the mice	Protection against zosteriform challenge[a]				Antibody responses[b]				CTL activities[c] (E:T = 100:1 EL-4/HSV)
	$10ID_{50}$	$50ID_{50}$	$100ID_{50}$	$500ID_{50}$	T IgG	IgG2a	IgG1	IgG2a/IgG1	
HSV-1 (KOS)	5/5	8/8	20/20	8/8	48648 ± 2863	12490 ± 3260	577 ± 88	21.6	36.85
pcgB DNA	4/5	4/8	9/20	2/8	2177 ± 88	796 ± 346	60 ± 25	13.3	9.21
pRJB-GM	5/5	8/8	14/20	4/8	3690 ± 17	531 ± 28	136 ± 3	3.9	23.62
Vectors	0/5	0/8	0/20	0/8	< 10	< 10	< 10	–	5.54

[a] 3–4 weeks old female BALB/c mice were injected with 2×10^6 HSV-1 (KOS) PFU, 90 μg of pcgB DNA + 50 μg of pRJB-GM, or 90 μg pcDNA I + 50 μg pRJB-GM, or 90 μg pcDNA I + 50 μg pRJB-c i.m. at day 0 and day 7. 10^4 PFU (10 ID_{50}), 5×10^4 PFU (50 ID_{50}), 10^5 PFU (100 ID_{50}), or 5×10^5 PFU (500 ID_{50}) of HSV-1.17 were zosteriform challenged to the mice at day 14. Skin lesions were recorded from day 21 to 28. The data shown in the table express the number of mice that did not develop skin lesions/total mice challenge.

[b] 3–4 weeks old female BALB/c mice were immunized as described above. At day 14, serum HSV-1 specific IgG and isotypes (IgG1, IgG2a) from the mice were analyzed by ELISA.

[c] 3–4 weeks old female C57BL/6 mice were immunized as described above. At day 14, spleen cells from the immunized mice were restimulated with u.v. inactivated HSV-1 (KOS) MOI = 1.5 in vitro for 5 days, followed by a Cr^{51} release assay. The data were expressed in specific lysis. Allogeneic target and mock controls had less than 5% specific lysis. Spontaneous releases were less than 15%.

in a 20 µl volume. Animals developed antibody responses detectable both in the serum and vaginal washings when measured 10 days after three intranasal injections. Inclusion of cholera toxin, a well known adjuvant for protein when administered to the alimentary tract (HOLMGREN et al. 1993), significantly enhanced the levels of immunity. The outcome in terms of levels of resistance induced to HSV challenge by the zosteriform as well as mucosal models is currently under investigation.

3 Mechanisms of Immunity Induced by Intramuscular Immunization with DNA

Sample animals in the immunization experiments with gB and ICP27 DNA discussed previously were killed at the time of challenge and serum and lymphoid cells were collected for immunological analysis. Details of these analyses have been published elsewhere (MANICKAN et al. 1995b 1995c). Briefly, both gB and ICP27 induced $CD4^+$ T cell responses as could be detected by lymphoproliferation and by antigen-induced cytokine production. The cytokines produced by DNA primed splenocytes were exclusively of the type 1 category, i.e., IFN-γ and interleukin (IL)-2 but not IL-4 or IL-10. Immunization with ICP27 DNA also induced a HSV-specific delayed-type hypersensitivity (DTH) response, normally also a correlate of a $CD4^+$ TH1 response (MOSSMANN et al. 1986). Whereas gB failed to induce CTLs at least in BALB/c mice, ICP27 DNA immunization did induce CTLs. Only gB immunization induced detectable antibodies and by ELISA these were predominantly of the Ig2a isotype of immunoglobulin, again a characteristic of CD4 helper T cell responses dominated by a Th1 subtype (NGUYEN et al. 1994). As mentioned previously in experiments using gB DNA along with GM-CSF, antibody responses were enhanced and interestingly the elevated species of Ig was principally of the IgG1 isotype.

To establish possible mechanisms of T cell immunity responsible for protection in the zosteriform model, adoptive transfer experiments were done using syngeneic athymic mice recipients. Using this model, protection against zosterification was shown to be a property of $CD4^+$ T cells and was not evident with $CD8^+$ T cells even from animals immunized with ICP27 DNA and which expressed $CD8^+$ immune activity in vitro. The possible role for antibody induced by DNA immunization was not measured in the zosteriform protection model. We can conclude from our DNA immunization experiments that an effective type 1 $CD4^+$ T cell response along with antibodies dominated by the IgG2a isotype was induced. Since the nature of the immune response elicited resembles that which occurs after live virus infection, and differs from that occurring after protein immunization (MANICKAN et al. 1995b), the DNA approach, once perfected, might represent a most valuable means of vaccination. This, however, remains to be seen.

4 Studies on Antigen Presentation with DNA Vaccines In Vitro

As indicated above, immunization with DNA is effective as a vaccine but exactly how immunity is achieved remains to be defined. The fact that purified "naked" DNA appears to be taken up and expressed by cells in vivo was surprising since in vitro few cells take up DNA unless a transfecting agent is used. For example, we have demonstrated using plasmid DNA encoding HSV proteins that when purified macrophages and DCs are exposed to naked DNA in vitro, only the former population express proteins, as measured either by western blotting or by their ability to act as APCs in primary CTL responses (ROUSE et al. 1994). However, if the same DNA is transfected into cells using the cationic lipid DOTAP then both macrophages (M_{ϕ}) and DC express proteins and act as APCs. In fact, DC are far superior than are M_{ϕ} at inducing primary CTL responses in naive T cell cultures. This is true both in comparisons of DC with M_{ϕ} to protein antigens given in the form of liposomes (NAIR et al. 1992a, b) as well as when transfected with DNA encoding herpesvirus proteins (Table 3). These CTL responses were shown to be antigen-specific, MHC-restricted and mediated by $CD8^+$ T cells.

Although the fact that naked DNA injected intramuscularly will induce MHC class I-restricted CTLs is widely accepted, exactly how this is accomplished remains in doubt. Using marker genes, expression of protein in muscle cells may occur for months after naked DNA injection (WOLFF et al. 1990). Although such cells do express class I MHC, it is not clear if they possess the appropriate processing machinery to generate target peptides. In addition, muscle cells certainly lack the costimulator molecules necessary for CTL induction. Indeed peptide presentation by muscle cells in the absence of costimulators might be expected to induce tolerance rather than activation of CTLs. Alternative explanations for CTL induction following intramuscular injection of naked DNA could include the fact that the pertinent APCs in muscles are the few cells of the DC series present at the time of injection, or inflammatory cells which rapidly invade as a sequel to damage

Table 3. Dendritic cells (DCs) pulsed with supernatant from macrophages (Mϕ) exposed to naked DNA induce primary antigen specific cytotoxic responses[a]

	EL4-gB		EL4	EMT6-gB
	12.5:1[b]	50:1[b]	50:1[b]	50:1[b]
pcDNA gB[c] + DC-T	1 ± 1	3 ± 1	2 ± 1	3 ± 1.5
Mϕ – pcDNA gB supernatant + T	2 ± 1.5	2 ± 2	1 ± 1	4 ± 2
Mϕ – pcDNA supernatant + DC-T	2 ± 1	2 ± 1	1.5 ± 1	3 ± 1
Mϕ – pcDNA gB supernatant + DC-T	21 ± 1	28 ± 3.5	2.5 ± 1	5 ± 2

[a] Mϕ (5×10^5 cells/ml) were treated with purified pcDNA gB or pcDNA (5 μg/ml) for 24 h in 96-well flat-bottom plates at 37° C. 100 μl of supernatant was harvested after 24 h, passed through a 0.45 μm filter and added to 100 μl of DC-T cell microcultures (R/S ratio 12.5:1) in 96-well U-bottom plates. CTL assay was done after 5 days at 37° C. Haplotypes of the cells used are indicated in parenthesis.
[b] Effector-target ratio.
[c] pcDNA gB – HSV-glycoprotein B (gB) cloned into pcDNA (Invitrogen).

associated with the injection. Another possibility is that the invading APCs ingest exogenous proteins or peptides released or regurgitated from the antigen-expressing muscle cells. Currently the idea that APCs acquire antigen exogenously and then process the material via the class I pathway is considered as improbable (GERMAIN 1994). However, there are an increasing number of examples whereby exogenous antigens can result in CTL induction and the general topic was briefly reviewed in a lucid article by BEVAN (1995).

Conceptually a number of possibilities exist. Firstly, phagocytes might possess a pathway to shunt protein from the phagosome into the cytosol where it can enter the MHC class I-associated pathway. This mechanism has been championed by Rock and colleagues (KOVACSOVICS-BANKOWSKI et al. 1993). Our group also provided support for this possibly using protein antigens delivered via pH sensitive liposomes (NAIR et al. 1992a, b). Secondly, phagocytes could digest ingested material in lysosomes, regurgitate peptides and these could load onto surface MHC class I of available cells. An example of this so-called vacuolar processing mechanism was described by PFEIFER et al. (1993). Thirdly, proteasome derived peptides might bind to a heat shock protein (HSP) and this complex binds to an unknown receptor on the APCs after which access to the cytosol of professional APCs occurs. Subsequently, the bound peptide is released from the HSP and is taken up by the nascent class I molecules. This mechanism has been advocated by SRIVASTAVA et al. (1993) as well as the by Rammensee group (ARNOLD et al. 1995). As discussed below, our results also provide evidence for the existence of such a mechanism (NAIR et al. 1996). The mechanism we describe accounts for cross-talk between M_ϕ and DC but it remains to be seen if a similar effect occurs between antigen-expressing myocytes and DCs and which could account for CTL induction by naked DNA vaccines.

In our experiments, naive splenic M_ϕ were exposed to DNA encoding HSV proteins and 24 h later culture fluids were collected and tested for immunogenicity in cultures of DC and naive T lymphocytes. After a 5 day period of in vitro culture, cells were tested for CTL activity. In such experiments, antigen-specific MHC-restricted CTL activity was generated. CTLs were not induced in T cell cultures without DCs nor in DC T cell cultures to which the naked DNA was added directly. Consequently, we interpret our observations to indicate that the M_ϕ cultures took up the DNA, expressed the protein, fragments of which were released outside the cell. This latter material was taken up by DCs, gained access to the appropriate processing pathway and triggered CTL precursors (CTL-p) which expanded and differentiated into CTLs.

We know that a minimum of 6 h was necessary for the M_ϕ expression and processing step to occur and that the immunogenic material released was larger than peptides which characteristically occupy the grooves of the T cell receptor. In fact, on the basis of studies aimed at neutralizing the immunogenicity using various antisera, the data indicate that the material is likely to consist of a peptide bound to a chaperon protein. Activity was inhibited for example with MAb to hsp 70 and gp 96 but not with MAb to hsp 90 or by some other control MAb (NAIR et al. 1996). We are currently attempting to determine if muscle cells taken from

DNA-immunized animals contain chaperon bound peptides that are immunogenic either in vitro or in vivo.

5 Epilogue

DNA immunization is currently enjoying great interest. This is probably justified particularly as a possible means of immunizing against agents currently refactory to control by vaccines. Should the approach prove efficacious against such agents, then the debut of DNA vaccines in 1993 represents a significant milestone along the course of events initiated by Jenner and Pasteur. Against HSV, an agent which currently lacks an acceptable vaccine, DNA immunization is efficacious at least when used prophylactically. This, however, cannot be considered as a noticeable milestone since a Pandora's box of approaches will protect the mouse against HSV induced disease when used prophylactically. Moreover, as currently used, DNA immunization still lacks the level of effectiveness attainable by other approaches. It could be that the use of adjuvant molecules will improve the situation and the observation that GM-CSF improves immunity is one step in that direction. Other more effective "tricks" are urgently required even for the use of DNA prophylactically against HSV. Currently, the value of DNA immunization as a potential therapeutic vaccine against HSV has not been assessed. Such studies are currently underway using the less than perfect guinea pig spontaneous reactivation model (Stanberry and Bernstein, personal communication). Herpes virologists anxiously await the results of their crucial experiments.

Acknowledgements. The last moment heroics of Paula Keaton are most appreciated. The authors' work is supported by AI 14981 and AI 33511.

References

Arnold D, Faath S, Rammensee HG, Schild H (1995) Cross-priming of minor histocompatibility antigen-specific cytotoxic T cells with the heat shock protein gp96. J Exp Med 182:885–889

Bevan MJ (1995) Antigen presentation to cytotoxic T lymphocytes in vivo. J Exp Med 182:639–641

Fynan EF, Webster RG, Fuller DH, Haynes JR, Santoro JC, Robinson HL (1993) DNA vaccines: protective immunizations by parenteral, mucosal, and gene-gun inoculations. Proc Natl Acad Sci USA 90:11478–11482

Germain RN (1994) MHC-dependent antigen processing and peptide presentation: providing ligands for T lymphocytes activation. Cell 76:287–299

Holmgren J, Lycke N, Czerknisky C (1993) Cholera toxin and cholera B subunit as oral mucosal adjuvant and antigen vector systems. Vaccine 11:1179–1184

Kovacsovics-Bankowski M, Clark K, Benacerraf B, Rock KL (1993) Efficient major histocompatibility complex class I presentation of exogenous antigen upon phagocytosis by macrophages. Proc Natl Acad Sci USA 90:4942–4946

Kriesel JD, Spruance SL, Daynes RA, Araneo B (1996) Nucleic acid vaccine encoding glycoprotein D2 protects mice from herpes simplex virus type 2 disease. J Infect Dis, In press

Kuklin N, Daheshia M, Karem K, Manickan E, Rouse BT (1997) Induction of mucosal immunity against herpes simplex virus by plasmid DNA immunization. J Virol 71:3138–3145

Kutinova L, Benda R, Kalos Z (1988) Placebo-controlled study with subunit herpes simplex virus vaccine in subjects suffering form frequent herpetic recurrences. Vaccine 6:223–228

Manickan E, Francotte M, Kuklin N, Dewerchin M, Molito C, Gheysen D, Slaoui M, Rouse BT (1995a) Vaccination with recombinant vaccinia viruses expressing ICP27 induces protective immunity against herpes simplex virus though $CD4^{+}$ $Th1^{+}$ T cells. J Virol 69:4711–4716

Manickan E, Rouse RJD, Yu Z, Wire WS, Rouse BT (1995b) Genetic immunization against herpes simplex virus protection is mediated by CD4 T lymphocytes. J Immunol 155:259–265

Manickan E, Yu Z, Rouse RJD, Wire WS, Rouse BT (1995c) Induction of protective immunity against herpes simplex virus with DNA encoding the immediate early protein ICP27. Viral Immunol 8:53–61

Mester JC, Rouse BT (1991) The mouse model and understanding immunity to herpes simplex virus. Rev Infect Dis 13:S935-945

Mossmann TR, Cherwinski H, Bond MW, Giedlin MA, Coffman RL (1986) Two types of murine helper T cell clones. J Immunol 136:2348–2357

Nair S, Rouse RJD, Bruce BD, Rouse BT (1996) Interaction between macrophages and dendritic cells involves a putative peptide chaperon. Submitted

Nair S, Zhou F, Huang L, Rouse BT (1992a) Class I restricted CTL recognition of a soluble protein delivered by liposomes containing lipophilic polylysines. J Immunol Methods 152:237–243

Nair S, Zhou F, Reddy R, Huang L, Rouse BT (1992b) Soluble proteins delivered to dendritic cells via pH-sensitive liposome induce primary cytotoxic T lymphocyte response in vitro. J Exp Med 175:609–612

Nguyen L, Knipe DM, Finberg RW (1994) Mechanisms of virus-induced Ig subclass shifts. J Immunol 152:478–484

Pfeifer JD, Wick MJ, Roberts RL, Findley K, Normark SJ, Harding CV (1993) Phagocytic processing of bacterial antigens for class I MHC presentation to T cells. Nature 361:153–177

Rouse RJD, Nair SK, Lydy SL, Bowen JC, Rouse BT (1994) Induction in vitro of primary cytotoxic T-lymphocyte responses with DNA encoding herpes simplex virus proteins. J Virol 68:5685–5689

Simmons A, Nash AA (1984) Zosteriform spread of herpes simplex virus as a model of recrudescence and its use to investigate the role of immune cells in prevention of recurrent diseases. J Virol 53:128–136

Simmons A, Tscharke DC (1992) Anti-CD8 impairs clearance of herpes simplex virus from the nervous system: implications for the fate of virally infected neurons. J Exp Med 175:1337–1344

Srivastava PK (1993) Peptide-binding heat shock protein in the endoplasmic reticulum: role in immune response to cancer and in antigen presentation. Adv Cancer Res 62:153–177

Stanberry LR (1996) Herpes immunization – on the threshold. J Eueop Acad Dermatol Venerol 7:120–128

Straus SE, Corey L, Burke RL, Savarese B, Barnum G, Krause PR, Kost RG, Meier JL, Sekulovich R, Adair SF (1984) Placebo-controlled trial of vaccination with recombinant glycoprotein D of herpes simplex virus type 2 for immunotherapy of genital herpes. Lancet 343:1460–1463

Witmer-Pack MD, Oliver W, Valinsky J, Schuler G, Steinman RM (1987) Granulocyte/macrophage colony-stimulating factor is essential for the viability and function of cultured murine epidermal Langerhans cells. J Exp Med 166:1484–1498

Wolff TA, Malone RW, Williams P, Chong W, Acsadi G, Jani A, Felgner PL (1990) Direct gene transfer into mouse muscle in vivo. Science 247:1465–1468

Xiang Z, Ertl HCJ (1995) Manipulation of the immune response to a plasmid-encoded viral antigen by coinoculation with plasmid expressing cytokines. Immunity 2:129–135

Humoral and Cellular Immune Responses to Herpes Simplex Virus-2 Glycoprotein D Generated by Facilitated DNA Immunization of Mice

C.J. PACHUK, R. ARNOLD, K. HEROLD, R.B. CICCARELLI, and T.J. HIGGINS

1 Introduction

Human herpetic infections have been described in ancient Greek and Roman times and are documented in the writings of both Hippocrates and Herodotus (NAHMIAS and DOWDLE 1968). The causative agents of these infections are viruses belonging to a large family of vertebrate pathogens known as the Herpesviridae. Classification of viruses as Herpesviridae is based on virion structure. Viral particles are comprised of an icosohedral capsid surrounded by an amorphous tegument. The core of the virion, which is enclosed by the capsid, contains the viral genome, a linear double-stranded DNA molecule ranging in size from 100 to several hundred kilobase pairs. The viral particle is surrounded by an outer lipid envelope into which the virus encoded glycoproteins are inserted (FIELDS et al. 1990). Several of the viral glycoproteins, including glycoprotein D (gD) of HSV (as discussed below), are essential for a productive viral infection in humans (HAY and RUYECHAN 1992).

At least seven species of herpesviruses have been shown to infect humans (FIELDS et al. 1990). These viruses include herpes simplex viruses 1 and 2 (HSV-1

Apollon, Inc., One Great Valley Parkway, Malvern, PA 19355, USA and HSV-2), human cytomegalovirus (CMV), varicella zoster virus, Epstein-Barr virus, human herpes virus 6 and human herpes virus 7.

Human herpes viruses are significant pathogens and have been shown to establish both lytic and latent infections (ROIZMAN and SEARS 1988; STEVENS 1975). Following a lytic infection, the virus is harbored in a latent state in the sensory ganglia. Reactivation from latency to lytic infection occurs throughout the lifetime of the infected host, although the underlying cause and mechanism of reactivation are still not understood. During episodes of reactivation, the host sheds infectious virus and may spread disease to susceptible persons (ROONEY et al. 1986).

Intervention of human herpes viral infections must involve new strategies which are designed to control infection, pathogenesis and reactivation. Although therapeutics, such as acyclovir (DORSKY and CRUMPACKER 1987), do exist for the modulation of lytic infection, they do not eliminate infection and therefore do not prevent future reactivation of disease. Over the last two decades, attempts have been made to develop prophylactic vaccines for HSV-1 and HSV-2, but clinical results have been disappointing (ROIZMAN 1991). However, a new prophylactic vaccine for varicella zoster has been recently approved (GERSHON 1995).

2 Herpes Simplex Virus-2 and Strategies for Vaccination

2.1 Herpes Simplex Virus-2: A Growing Health Problem

Herpes simplex virus-2 is a sexually transmitted disease that affects approximately 50 000 000 people in the United States alone. The incidence of HSV-2 genital infections continues to rise at a high rate and is attributed to several factors: (1) the high frequency of undiagnosed disease, (2) the chronicity of infection and (3) subclinical episodes of viral shedding. Furthermore, establishment of HSV-2 infection is not prevented by previous infection with HSV-1 (NAHMIAS et al. 1990).

2.2 Vaccine Strategies for Herpes Simplex Virus-2

A successful vaccine for the prevention of genital herpes is one that would prevent primary HSV-2 infection and therefore preclude the colonization of sensory ganglia by latent viruses. There are several considerations in the design of such a vaccine. The vaccine should elicit both humoral and cellular immune responses, and a subset of these responses should be neutralizing. The choice of a viral immunogen(s) and type of immunization is therefore tantamount to developing an efficacious vaccine. Recent clinical studies have focused on intramuscular vaccination with a combination of viral glycoproteins (gB and/or gD) with various adjuvants (BURKE 1992).

3 Facilitated DNA Vaccines

We have focused our efforts on the development of a DNA vaccine for HSV-2. Facilitated DNA vaccines involve the direct parenteral injection of plasmid DNA encoding viral, bacterial or tumor genes with a facilitator of DNA uptake and expression, such as bupivacaine (Wang et al. 1993). The direct uptake and expression of DNA in vivo allows the presentation of viral, bacterial or tumor peptides in the context of both class I and class II MHC antigens. Previous studies have shown that parenteral injection of an HIV-1-gp160 expression vector in the presence of the facilitating agent bupivicaine induces both a humoral and cytotoxic T cell CTL response to gp160 in vaccinated rodents and nonhuman primates, including chimpanzees (Wang et al. 1993). Recent studies have shown that cynmologous monkeys vaccinated with this vector were able to control viral load following challenge with 50 animal infectious doses of recombinant SHIV (Wang et al. 1995). In addition, DNA immunization has also been shown to induce class I restricted CTL responsiveness in a mouse strain that is an antigen specific nonresponder when immunized by methods that induced CTL activity in other mouse strains (Schirmbeck et al. 1995).

4 Herpes Simplex Virus Glycoprotein D as an Immunogen

Glycoprotein D is present on both the HSV viral envelope and the surface of HSV-infected cells (Hay and Ruyechan 1992) and is a target for neutralizing antibody (Highlander et al. 1987) and cellular immune responses (Blacklaws et al. 1987). Vaccination with recombinant gD has been shown to be protective against viral challenge in several animal systems and to elicit long-term immune responses in primates (Mishkin et al. 1991). Vaccines using recombinant glycoprotein D or a combination of truncated glycoproteins B and D (Burke 1992) have been evaluated in human clinical trials. Due to the broad-based immune response induced by facilitated DNA immunization, it is reasonable to explore this technology using vectors expressing gD for the development of prophylactic and therapeutic vaccines for HSV-2.

5 Results and Discussion

5.1 Construction of the Expression Vector pAPL-gD2

The gD gene of an HSV-2 clinical isolate, HSV-2 (strain 12), was cloned into the expression vector pAPL-1. The resultant construct, pAPL-gD2 (Fig. 1), contains a

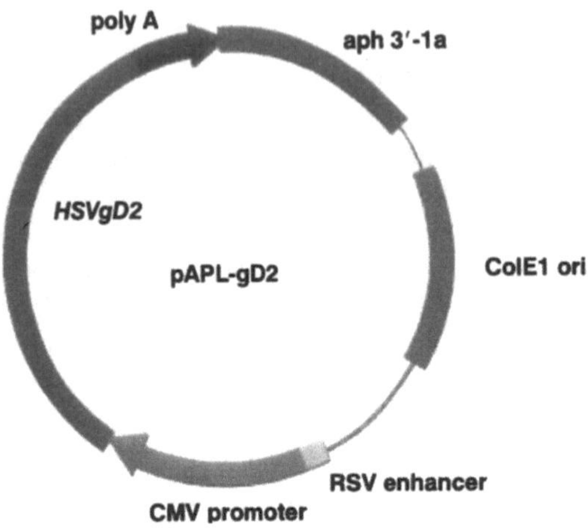

Fig. 1. Functional map of pAPL-gD2. The HSV-2 strain 12 gD gene was cloned under the transcriptional control of the cytomegalovirus (CMV) promoter and Rous sarcoma virus (RSV) enhancer element. An SV40 derived polyadenylation signal is provided at the 3′ end of the insert. The plasmid contains a *Col*E1 origin of replication and an *aph* 3′-1a gene encoding aminoglyocside resistance for selection of plasmid containing bacteria

2 kb fragment that encodes the entire gD gene and several hundred nucleotides of 5′ and 3′ flanking sequence, placed under the transcriptional control of the human CMV promoter and the Rous sarcoma virus (RSV) enhancer element. An SV40 polyadenylation signal is supplied by the expression vector.

The sequence of the HSV-2 (strain 12) gene is similar to that described for HSV-2 strain G (Watson 1983) (Table 1). Although pAPL-gD2 carries three nucleotide changes in the gene sequence, only one results in an amino acid codon change when compared to that of strain G.

5.2 Analysis of In Vitro Expression

Expression of gD from pAPL-gD2 was assayed in vitro in a human rhabdomyosarcoma (RD) cell line. RD cells were transfected with pAPL-gD2 or pAPL-1 (the vector control) and assayed at 48 h post-transfection by immunofluorescence and immunoblot analysis.

Transfected RD cells were analyzed by indirect immunofluorescence using an anti-HSVgD monoclonal antibody, Dl-6 (Isola et al. 1985), as the primary anti-

Table 1. Sequence comparison of the glycoprotein D (gD) gene of HSV-2 strains 12 and G

Nucleotide position	Nucleotide change		Amino acid change	
	Strain 12	Strain G	Strain 12	Strain G
733	C	T	Leu	Leu
735	C	A	Leu	Leu
1058	T	C	Val	Ala

The positions of the three nucleotide changes seen between the HSV-2 strains 12 and G gD genes are shown relative to the initiation codon, ATG, where nucleotide A is position 1. The single amino acid change of Ala to Val is also indicated.

body. Fluorescein isothiocyanate (FITC)-labeled anti-mouse IgG was used as the secondary antibody. Only cells transfected with pAPL-gD2 and fixed in methanol/acetone prior to immunostaining exhibited bright cytoplasmic fluorescence (Fig. 2). Immunostaining of live cells was also performed to determine if gD was expressed on the surface of cells transfected with pAPL-gD2. Since gD is a membrane protein associated with both the viral envelope and the infected cell surface, it was anticipated that expression of gD in transfected cells would result in a population of gD molecules localized to the cell surface. The immunostaining results indicate that gD is expressed on the surface of only those cells transfected with pAPL-gD2 (Fig. 3).

The molecular weight of gD protein present in the pAPL-gD2 transfected cell lysates was determined by immunoblot analysis. A single molecular weight species in the pAPL-gD2 transfected cell lysates reacted with the DI-6 monoclonal antibody (Fig. 4). The apparent molecular weight of this species is approximately 50–53 kDa and is larger than the predicted molecular weight for gD of 40.5 kDa. The increase in molecular weight is probably due to the known glycosylation of gD, which contains both N-linked and O-linked sugar residues (SERAFINI-CESSI et al. 1988). When transfections were carried out in the presence of tunicamycin, an inhibitor of N-linked glycosylation, the molecular weight of gD in the cell lysates was shifted to 41–43 kDa (data not shown). This result is consistent with the interpretation that pAPL-gD2 transfected cells synthesize glycosylated gD molecules. In summary, the in vitro expression analysis indicates that pAPL-gD2 directs the synthesis, in RD cells, of an approximately 55 kDa, glycosylated, cell surface protein, similar to the gD that is synthesized in HSV-2 infected cells.

A B

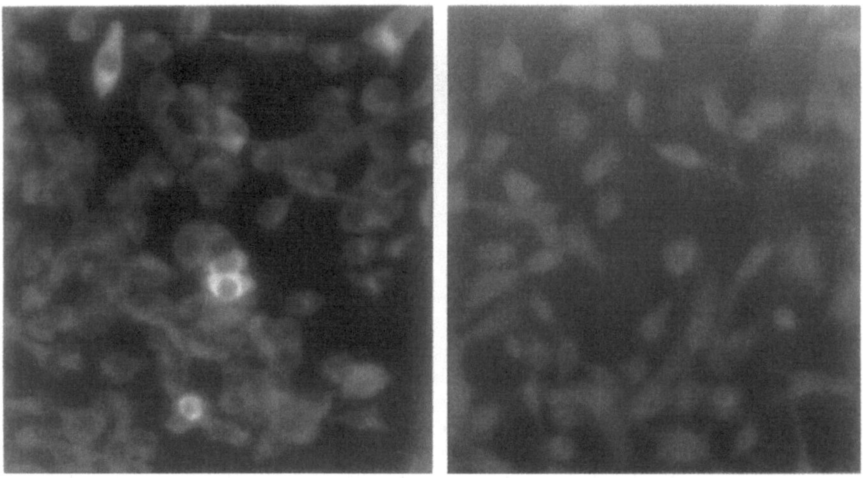

Fig. 2A, B. Analysis of HSV-2 glycoprotein D (gD) expression by immunofluorescent staining. Rhabdomyosarcoma cells were transfected with pAPL-gD2 (**A**) or pAPL-1 (**B**) and fixed in 50% methanol/50% acetone prior to indirect immunostaining. The primary antibody used in this experiment was an anti-HSVgD monoclonal antibody, and the secondary antibody was an FITC-labeled anti-mouse antibody

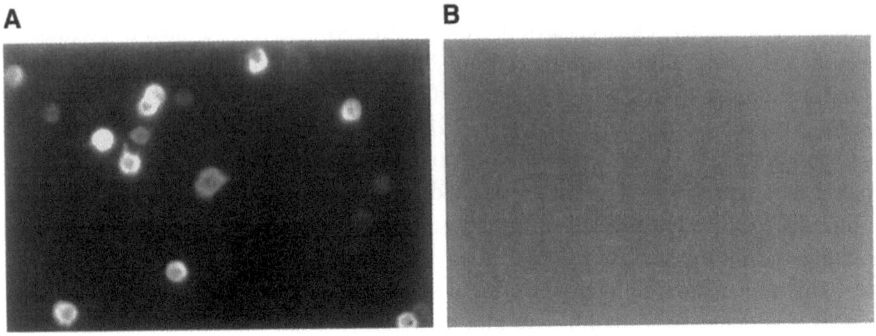

Fig. 3A, B. Expression of gD on the surface of transfected cells. Rhabdomyosarcoma cells were transfected with pAPL-gD2 (**A**) or pAPL-1 (**B**) and subjected to live cell immunostaining. The primary antibody used in this experiment was an anti-HSV-2gD monoclonal antibody, and the secondary antibody was an FITC-labeled anti-mouse antibody

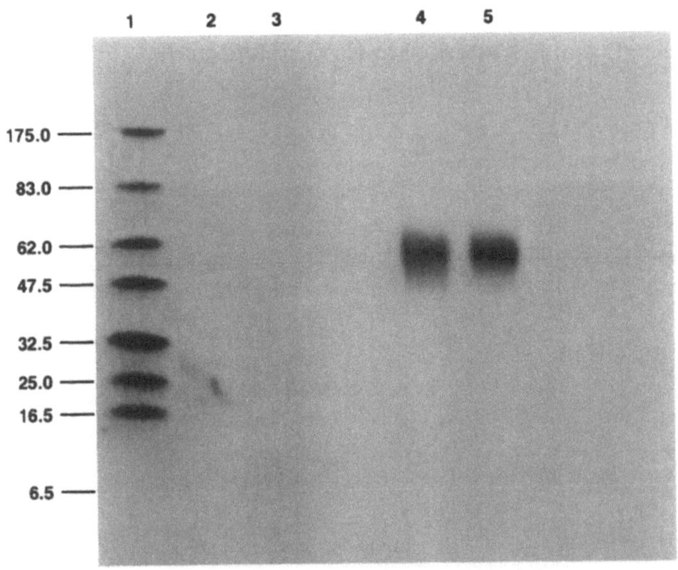

Fig. 4. Detection of a 55 kDa, HSV-2 glycoprotein D (gD) specific protein in pAPL-gD2 transfected rhabdomyosarcoma cells. Rhabdomyosarcoma cells were transfected with pAPL-gD2 (*lanes 4, 5*) or control vector (*lanes 2, 3*). At 48 h post-transfection, cell lysates were harvested and subjected to SDS-PAGE. Following electrophoresis, protein was transferred to nitrocellulose paper. HSV-2gD sequences were identified by incubating blots with the monoclonal antibody D1-6. An anti-mouse IgG polyclonal antibody conjugated to alkaline phosphatase was used as the secondary antibody. Blots were developed by incubating in the substrate 5′-bromo – 4 chloro – 3 indolyl phosphate/nitro blue tetrazolium. Molecular weights in kilodaltons of protein size standards (*lane 1*) are indicated

5.3 Production and Purification of pAPL-gD2

In order to produce material for the evaluation of immune responses in vaccinated mice, pAPL-gD2 plasmid DNA was made by fermentative growth of an *E.coli* K12 strain, DH10B, that was previously transformed with pAPL-gD2. The pAPL-gD2 supercoiled DNA was purified by a multi-step chromatography procedure and was formulated directly with the local anesthetic bupivacaine. A control vector containing the *E. coli* β-galactosidase gene (pAPL-Bgal) was also prepared and formulated in the same manner.

5.4 Seroconversion in Immunized Mice

Mice injected with bupivacaine formulated pAPL-gD2 DNA gave consistent and high levels of seroconversion. The results of one experiment are shown in Fig. 5. All five mice seroconverted by the first evaluation point (2 weeks) and their titer continued to rise following a booster injection at 30 days. In another group, all five mice seroconverted following a single injection of 100 μg of bupivacaine formulated pAPL-gD2 and reached maximal antibody production between day 14 and day 42. Antibody levels in this group were still at or near maximal 215 days following a single injection.

It was possible, although unlikely, that the seroconversion observed upon injection of formulated pAPL-gD2 was due to a nonspecific elevation of serum immunoglobulin. To address this point, day 42 serum from a pAPL-gD2 injected animal was titered in an ELISA assay on plate bound β-galactosidase. Figure 6 shows that antiserum from these animals does not bind the β-galactosidase antigen. The reciprocal experiment was also performed and produced identical results, i.e., antisera from pAPL-Bgal immunized animals had insignificant binding to plate bound gD2. In both of these experiments, sera showed an A_{450} of ≥2 with the appropriate antigen. This experiment was repeated with several animals with similar results.

Injection of bupivacaine formulated DNA is a very efficient method of immunization, as shown by the representative titration data in Fig. 7. For this experiment, groups of five mice were injected with bupivacaine formulated pAPL-gD2 at doses of either 100 μg, 10 μg or 1 μg of DNA. As before, all five mice in the 100 μg group seroconverted by day 14 and reached similar high levels of antibody production by day 42. All five mice in the 10 μg DNA group also seroconverted by day 14, and two of the five had already reached antibody levels comparable to the 100 μg DNA group by day 42. Therefore, maximal seroconversion can be obtained with as low as 10 μg of DNA per mouse. Furthermore, the data in Fig. 7 also show that 1 μg of formulated pAPL-gD2 produced seroconversion in injected mice. In the 1 μg DNA group, four of five animals seroconverted after a single 1 μg DNA injection, and all five had measurable antibody to gD2 by day 42. These data indicate that the bupivacaine formulated pAPL-gD2 stimulates a rapid, consistent seroconversion in mice which can be observed at 1 μg of DNA.

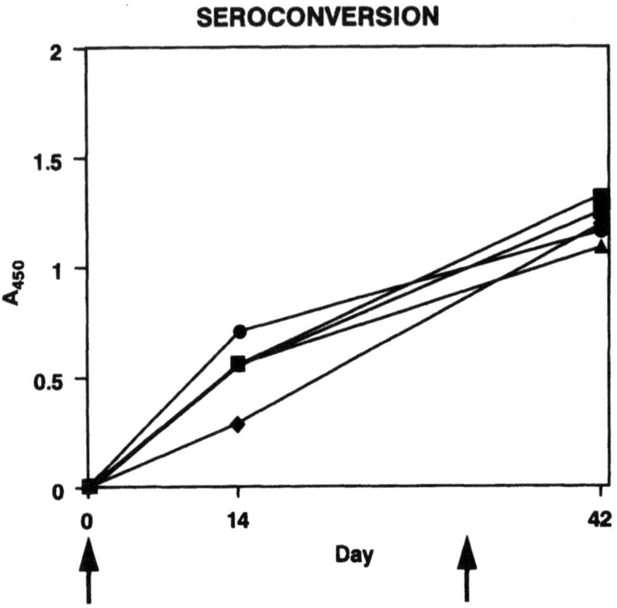

Fig. 5. Seroconversion to pAPL-gD2. Five Balb/c mice were inoculated with 100 µg pAPL-gD2 formulated with bupivacaine in the quadriceps and boosted with the same dose 1 month later. Serum was collected and analyzed by ELISA using 20 ng of recombinant gD2 per well. Pre-bleed values have been subtracted. *Arrows* represent day of injection

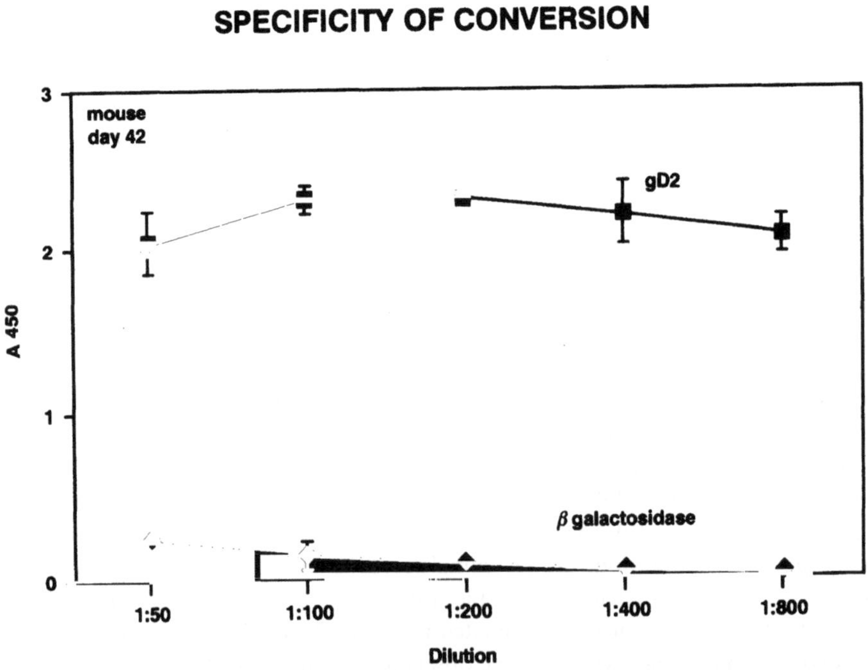

Fig. 6. Specificity of glycoprotein D2 (gD2) antiserum. Serial dilutions of day 42 antiserum from an animal injected with pAPL-gD2, as described in Fig. 5, were titered on ELISA plates containing either recombinant gD2 or β-galactosidase. Data are presented as for Fig. 5; *closed squares*, gD2; *closed diamonds*, β-galactosidase

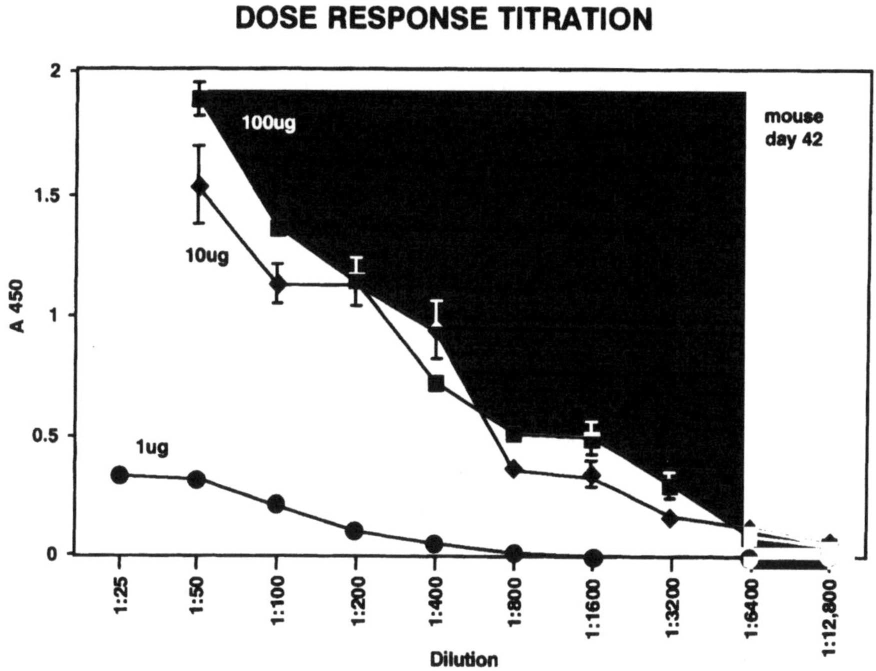

Fig. 7. pAPL-gD2 dose response. Groups of five Balb/c mice were injected with different amounts of pAPL-gD2 formulated with bupivacaine and boosted at 1 month. Day 42 antiserum was serially diluted and assayed by ELISA on gD2 protein. *closed squares*, 100 μg dose; *closed diamonds*, 10 μg dose; *closed circles*, 1 μg dose

Table 2. Spleen cell lymphoproliferation following immunization with bupivacaine facilitated pAPL-gD2. Mice were injected with either 100 μg, 10 μg or 1 μg pAPL-gD2 formulated with bupivacaine in the quadricept muscle and similarly boosted three times at one month intervals. Spleens were harvested one month after the last boost. Wells containing 10^5 cells were stimulated with either 20 ng/well of recombinant gD2 antigen or 50 ng of recombinant HIV-1 gp120 (control) and pulsed with 3H-thymidine 18 hours before harvest on day five. Data is presented as stimulation index (S.I.; experimental/background) and elevation of cpm over background (Δ cpm) ± standard deviation

Facilitated DNA	Antigen	SI	ΔCPM
100 μg	gD2	13.7 ± 6.5	19,047 ± 4,813
	gp120	1.1 ± 0.3	293 ± 518
10 μg	gD2	9.0 ± 3.0	2,549 ± 1,556
	gp120	1.0 ± 0.2	−12 ± 29
1 μg	gD2	4.3 ± 2.9	2,256 ± 1,307
	gp120	1.0 ± 0.2	−104 ± 163

SI, Stimulation index.

5.5 Lymphoproliferation in Immunized Mice

One of the major advantages of DNA vaccines is the stimulation of both the humoral and cellular immune responses. One measure of the cell mediated response is proliferation of spleen lymphocytes to the immunizing antigen. Spleens from the facilitated plasmid dose response experiment were harvested at the end of the experiment and assayed for lymphoproliferation. The results of this assay are shown in Table 2. A significant response to the gD2 antigen, as determined by both stimulation index (S.I.) and the elevation of cpm over background, is observed in all mice in both the 100 μg and 10 μg groups. At least half of the mice in the 1 μg group showed a significant proliferative response. Results similar to those shown in Table 2 have been a consistent finding in these studies. The lymphoproliferation results indicate that, like the humoral response, the cell mediated arm of the immune response is highly responsive to facilitated DNA immunization.

The level of seroconversion and lymphoproliferation observed in our studies using facilitated DNA vaccines compare quite favorably to the studies which used gD2 protein and alum as an adjuvant (Landolfi et al. 1993). Additional experiments exploring the proliferative response, CTL response and protection generated by facilitated APL-gD2 vaccination are in progress using mice, rabbits and non-human primates.

Acknowledgements. The authors wish to thank Dr. C. Satishchandran for critical review of the manuscript, and Teresa Moore for assistance in manuscript preparation.

References

Blacklaws BA, Nash AA, Darby G (1987) Specificity of immune response of mice to herpes simplex virus glycoproteins B and D constitutively expressed on L cell lines. J Gen Virol 68:1103–1114

Burke RL (1992) Contemporary approaches to vaccination against herpes simplex virus. Curr Top Micro and Immun 179:137–158

Dorsky DI, Crumpacker CS (1987) Drugs five years later: acyclovir. Ann Intern Med 107:859–874

Fields BN, Knipe D, Chanock RM, Hirsch MS, Melnick JL, Monath TP, Roizman B (1990) Virology (2nd ed), Raven, New York

Gershon AA (1995) Varicella-zoster virus: prospects for control. Adv in Pediatric Inf Dis 10:93

Hay J, Ruyechan WT (1992) Regulation of herpes simplex virus type 1 gene expression. Curr Top Micro Immun 179:1–14

Highlander SL, Sutherland SL, Gage PJ, Johnson DC, Levine M, Glorioso JC (1987) Neutralizing monoclonal antibodies specific for herpes simplex virus glycoprotein D inhibit virus penetration. J Virol 61:3356–3364

Isola VJ, Eisenberg RJ, Siebert GR, Heilman CJ, Wilcox WC and Cohen GH (1985) Fine mapping of antigenic site II of herpes simplex virus type 1 glycoprotein D. J Virol 63:2325–2334

Landolfi V, Zarley CD, Abramovitz AS, Figueroa N, Wu SL, Blasiak M, Ishizaka ST, Mishkin EM (1993) Baculovirus-expressed herpes simplex virus type 2 glycoprotein D is immunogenic and protective against lethal HSV challenge. Vaccine 11:407–414

Mishkin EM, Fahey JR, Kino Y, Klein RJ, Abramovitz AS, Mento SJ (1991) Native herpes simplex virus glycoprotein D vaccine: Immunogenicity and protection in animal models. Vaccine 9:147–153

Nahmias AJ, Dowdle WR (1968) Antigenic and biologic differences in herpesvirus hominis. Prog Med Virol 10:110–159

Nahmias AJ, Lee FK, Beckman-Nahmias S (1990) Sero-epidemiological and -sociological patterns of herpes simplex virus infection in the world. Scand J Infect Dis (Supp) 69:19–36

Roizman B (1991) Introduction: objectives of herpes simplex virus vaccines seen from a historical perspective. Rev Infect Dis (Supp 11) S892–S894
Roizman B, Sears A (1988) Inquiring into mechanisms of herpes simplex virus latency. Am Rev Microbiol 41:543–571
Rooney JJ, Felser JM, Ostrove JM, Straus SE (1986) Acquisition of genital herpes from an asymptomatic sexual partner. N Eng J Med 314:1561–1564
Schirmbeck R, Böhm W, Ando K, Chisari FV, Reimann J (1995) Nucleic acid vaccination primes. Hepatitis B virus surface antigen-specific cytotoxic T lymphocytes in nonresponder mice. J Virol 69:5929–5934
Serafini-Cessi F, Dall'Olio F, Malagolini N, Pereira L, Campadelli-Fiume G (1988) Comparative study on O-oligosaccharides of glycoprotein D of herpes simplex virus types 1 and 2. J Gen Virol 69:869–877
Stevens JG (1975) Latent herpes simplex virus and the nervous system. Curr Top Immunol 70:31–50
Wang B, Boyer JD, Ugen KE, Srikantan V, Ayyaroo V, Agadjanyan MG, Williams WV, Newman M, Coney L, Carrano R, Weiner DB (1995) Nucleic acid based immunization against HIV-1: induction of protective in vivo immune responses. AIDS 9 (Supp A):S159–S170
Wang B, Ugen KE, Srikantan V, Agadjanyan MG, Dang K, Refaeli Y, Sato AI, Boyer J, Williams WV, Weiner D (1993) Gene inoculation generates immune responses against human immunodeficiency virus type. 1 Proc Natl Acad Sci 90:4156–4160
Watson R (1983) DNA sequence of the herpes simplex virus type 2 glycoprotein D gene. Gene 26:307–312

Nucleic Acid Vaccines: Veterinary Applications

L.A. Babiuk, P.J. Lewis, S. van Drunen Little-van den Hurk, S. Tikoo, and X. Liang

1 Introduction

Few would disagree that immunization to prevent infectious diseases has significantly impacted the economics of livestock and poultry production globally. Unfortunately, even with these successes, there is a need to extend the impact of such vaccination programs since economic losses continue to plaque producers and in some instances limit the areas of livestock production or reduce the economic return to producers by limiting the export of genetic stock. To increase animal productivity, infectious diseases have been the primary targets for vaccination, although more recently, ectoparasites are also becoming a target for immunization.

Most currently licensed vaccines are produced by conventional technologies which have not changed significantly from those used by Jenner or Pasteur 200 and 100 years ago, respectively. Due to the inherent disadvantages of these conventional vaccines, novel strategies are being developed to produce a new generation of vaccines with improved efficacy, ease of administration and fewer side reactions. These new generation vaccines have been made possible by recent advances in molecular biology and our understanding of virulence genes, identification of specific proteins or glycoproteins involved in inducing protective immunity, as well as an appreciation of host-pathogen interactions and host immune responses to various pathogens. This is especially important for those pathogens which cannot

Veterinary Infectious Disease Organization, 120 Veterinary Road, Saskatoon, Saskatchewan S7N 5E3, Canada be cultured easily in vitro. The first vaccine produced by this technology, for any species, was a vaccine designed to control hepatitis B. In this case, the hepatitis B surface antigen was produced in yeast (Valenzuela et al. 1982). This was followed shortly thereafter by the production of the world's first genetically engineered licensed subunit vaccine for animals when the *Pasteurella hemolytica* leukotoxin was cloned, expressed in *E. coli* and shown to be effective in reducing respiratory disease in cattle (Potter et al. 1990). Subsequently, the licensing of a live, gene deleted vaccine for pseudorabies virus not only provided for a safe live vaccine, but also allowed for the implementation of a immunization-eradication program to eliminate pseudorabies virus (Kit et al. 1987). These initial successes have lead to the licensing of a number of live recombinant vaccines such as vaccinia rabies glycoprotein recombinants (Pastoret et al. 1988). However, even as this new generation of vaccines is being developed and accepted by regulatory agencies, commercial companies and the general public, a new approach to immunization has emerged which has the potential to complement these new generation vaccines and in some cases supplant them completely.

The third generation of vaccines is based on the observations that naked nucleic acids, when introduced into cells, can function and express sufficient proteins in vivo to evoke an immune response (Tang et al. 1992; Cox et al. 1993; Ulmer et al. 1993). Following these initial observations, the phenomenon of genetic immunization has been confirmed with genes encoding for viral, bacterial, parasitic, and tumor antigens. Furthermore, this phenomenon appears to work in many species including those of veterinary importance (Williams et al. 1991; Tang et al. 1992; Cox et al. 1993; Davis et al. 1993; Donnelly et al. 1993; Fynan et al. 1993; Robinson et al. 1993b; Hoffmann et al. 1994; Lowrie et al. 1994; Xiang et al. 1994; Xu et al. 1994). The present review will summarize the practical problems associated with vaccine development for veterinary species as well as the application of genetic immunization to overcome some of these problems.

2 Impediments to Developing the Ideal Vaccine

The ideal vaccine for animals would be one that provides greater than 90% efficacy following a single administration of the vaccine. Furthermore, this protection should occur rapidly, within 2 weeks of administration, and should be of long duration such that animals would not need to be re-immunized. This would not only reduce the cost of repeated immunization but also reduce the level of pathogen circulation in the environment. Even though it may be possible to achieve success under experimental conditions, consideration must be given to the different management conditions employed in different parts of the world. Furthermore, epidemiology and pathogenesis of the specific agent will influence the type of vaccine that will be effective as well as the effectiveness of different vaccination regimes. For example, in cattle and pigs, enteric infections caused by rotavirus, coronavirus and *E. coli* occur in the first few weeks of life. Since it is not possible to induce immunity within 2–3 days of birth, with any vaccine, the main focus is to immunize the mother who then transfers antibody in her milk to the offspring. This antibody coats or bathes the gastrointestinal tract of the young and prevents infection. Thus, the ideal vaccine for this group of pathogens would be one that is administered to the dam, resulting in high levels of antibody in the dams milk for 3–4 weeks (Babiuk et al. 1984). To achieve this, immunization regimes will be required to stimulate high levels of antibody in the milk of the mothers to protect the offspring. Unfortunately, with early weaning practices in pigs, animals are removed from the mothers while they are still susceptible to infection. Thus, immunization regimes will need to be developed whereby passive immunity provides early protection while active immunity is being developed. Unfortunately, passively acquired antibody also interferes with active immunization by conventional vaccines. Thus, novel approaches to overcome interference by passive antibody will need to be developed. Recent studies have demonstrated that genetic immunization has the potential to overcome interference by passive antibody.

In poultry, infection occurs shortly after hatching. Thus, producers have immunized layers, who then transfer antibody to the eggs thus protecting the chicks from infections early in their lives. This is especially important for diseases such as infectious bursal disease, which during the first few weeks of life destroys the bursa of Fabricius and subsquently reduces the bird's ability to develop immunity to many pathogens because the number of B cells is reduced. An alternate approach to provide early immunity is to immunize day-old chicks. Unfortunately, some diseases occur before the chick develops significant levels of immunity. To try and induce active immunity early, Embrix (Durham, NC) has developed a method to introduce vaccines into chicks prior to hatching (in ovo vaccination). This new egg inoculation technique also reduces the amount of handling required for mass immunization of poultry.

In the case of bovine respiratory disease, management systems vary around the world with economical losses often occurring following weaning, movement and confinement of animals. This type of management system not only stresses the animals, thereby making them more susceptible to infections, but also provides an excellent opportunity for simultaneous exposure to a variety of different pathogens (Babiuk and Campos 1992). In many cases these pathogens act synergistically to increase the severity of respiratory infection. Vaccination at entering the feedlots does not provide sufficient time for the development of immunity and disease prevention. Reduction in infection and economic losses could be achieved if pre-shipping vaccination regimes were adopted. If vaccination could be implemented early, this would reduce infections, but unfortunately animal husbandry practices prevent this type of approach from being implemented.

In many cases it is unsatisfactory to introduce vaccines by intramuscular or subcutaneous routes since this requires needle injection. In commercial settings the rather large number of animals that are immunized at any one time, which can involve restraint difficulties, can lead to pathologies such as: injection site necrosis, focal adjuvant or bacterial mediated myositis, abcessation, and broken needles which may remain deposited in the muscle of the animals. This affects meat quality and safety. Thus, it is imperative that better vaccines and delivery systems are developed which reduce these unwanted side effects. Delivery of vaccines orally or intranasally should remove some of these unwanted side effects. A further advantage of this type of delivery is that most infections enter via a mucosal surface, thus, immunity at these sites would prevent initiation of the infection. In poultry, the sheer number of animals involved in any single enterprise favors delivering the antigen in water, by aerosols, the food or in ovo. In this species the ideal vaccine would be one that takes into consideration ease of administration while ensuring that mucosal protection is induced.

Another very important consideration in veterinary medicine is the cost. This is especially important for poultry and in developing countries. Vaccines for poultry must be produced for pennies per dose. This is in contrast to vaccines for other mammalian species which may sell for \$0.50–\$1.00. In humans similar vaccines could sell for \$10.00–\$30.00. This economic difference clearly indicates the impediments to developing effective vaccines for some veterinary species. Thermal stability is required in most situations where refrigeration is not possible to maintain the vaccines from manufacturing to administration. This is especially important in developing countries where a cold chain is not always guaranteed or for those vaccines which are administered in bait and may be present in the environment for extended periods of time before infection (Pastoret et al. 1988).

In addition to developing mucosal immunity the vaccine should develop a balanced immune response inducing both humoral and cellular immunity. In all instances, vaccines need to be genetically stable with no chance of reversion to virulence in causing clinical signs either in the vaccinated animals or in unvaccinated contacts. Immunosuppression or interference with immunity to other vaccines given simultaneously has been shown to occur, but should be minimal or nonexistent (Harland et al. 1992; Kerlin et al. 1992; Inokuma et al. 1993; Theodos et al. 1993; Wikel et al. 1994). Since many vaccines are administered simultaneously, vaccine architecture and compatibility are becoming very important considerations (Morein et al. 1987; Hughes et al. 1991; Hughes et al. 1992b; East et al. 1993). Although at present many adjuvants contain mineral oils, these will need to be replaced in the future since these oils are nonmetabolizable and leave residues at the injection site. All of these factors must be considered in developing vaccines for veterinary species.

3 Novel Approaches to Immunization

Due to the need for better vaccines for many diseases, new approaches are being developed for vaccine production and delivery. These approaches are based on the observation that pathogens possess a variety of different proteins, some of which are important in inducing protective immunity whereas others are of no relevance in protection and can actually be detrimental by virtue of their ability to induce immunosuppression or even enhance infectivity (PEDERSEN et al. 1986; NICK et al. 1990; OLSEN et al. 1992; HARLAND et al. 1992; KERLINE et al. 1992; INOKUMA et al. 1993; WIKEL et al. 1994). The thrust over the past decade has been to develop either subunit vaccines containing only the most critical proteins involved in inducing protection or live attenuated vaccines by deleting specific genes involved in virulence with a low probability of back mutation and reversion to virulence (KIT et al. 1985). This should ensure safer vaccines that can be delivered by the oral or mucosal routes and stimulate both cellular and humoral immunity at mucosal sites, thereby blocking the establishment of an infection. Even more exciting is the possibility to use these genetically engineered vectors to carry multiple antigens from other pathogens, thereby providing the opportunity for immunizing animals against a variety of pathogens with a single vaccine (PASTORET et al. 1988; YILMA et al. 1988; ZIJL et al. 1991; WHITTON et al. 1993). This approach has many advantages over multiple vaccinations or combination vaccines in which one pathogen may interfere with immune responses to the other. One of the best examples of such a genetically engineered vaccines is vaccinia rabies glycoprotein recombinants, which are used to control rabies in wildlife (PASTORET et al. 1988).

Recent progress in molecular biology, protein engineering and immunity has had a very significant impact on our ability to identify protective components of many pathogens including those which stimulate T and B cell responses. This reductionist approach has proven to be successful for both viral and bacterial pathogens and more recently with "concealed" antigens of ectoparasites (OPDEBEECK 1994). However, these studies have also indicated that the form of antigen and its formulation have a significant influence on vaccine efficacy and on the duration of the immunity (VAN DRUNEN LITTLE-VAN DEN HURK et al. 1993a). This is especially important for those pathogens in which cell-mediated or mucosal immunity is critical. In viral infections, the major targets for immunity are often surface glycoproteins. Since glycosylation and conformational properties of the glycoprotein are critical for inducing the appropriate immune response, they must be produced in mammalian cells. Presently this is relatively expensive and even if the protein is produced in mammalian cells the immune response to exogenously delivered antigen is not the same as to endogenously produced antigens.

Since the form of antigen is critical for inducing the appropriate immune responses as well as for long-term immunity in animals, we attempted to determine whether immunity induced by genetic immunization would function in livestock. This hypothesis was based on earlier studies in which transformation of cells in vitro occurred following injection of purified polyomavirus naked DNA (ATANASIU 1962). These studies were further extended by WOLFF (1990), who clearly indicated that injection of plasmid DNA into muscle cells of mice resulted in expression of foreign genes in the cells for extended periods of time. This was latter confirmed by TANG et al.(1992) and numerous other workers using mouse model systems. These studies also clearly indicated that both cell mediated immunity and humoral immunity were induced following in vivo introduction of plasmid DNA.

As a result of expression of these genes in vivo, immunity to foreign proteins develops and may remain for extended periods of time (ULMER et al. 1993). Since uptake and expression in vivo has been observed with various genes in different species such as fish, chickens, dogs, cats, monkeys and cows it appears to be an universal phenomenon which would have application in all veterinary species (WILLIAMS et al. 1991; WEISS et al. 1992; DONNELLY et al. 1993; FYNAN et al. 1993; ROBINSON et al. 1993b; HOFFMANN et al. 1994; LOWRIE et al. 1994; XIANG et al. 1994; XU et al. 1994; ULMER et al. 1994a).

In an attempt to demonstrate the utility of this approach in veterinary species our laboratory has focused on developing vaccines against viruses for cattle (COX et al. 1993). More specifically we have focused on bovine herpesvirus-1 and bovine rotavirus (BABIUK et al. 1996). The following section will describe some of our observations and describe possible extensions of our studies to other pathogens of veterinary interest.

Vaccines to bovine herpesviruses can serve as a model for herpesvirus infections in other species as well as help reduce significant losses in the cattle industry due to infection by this virus (VAN DRUNEN LITTEL-VAN DEN HURK et al. 1993b; TIKOO et al. 1995). Since the surface glycoproteins are critical for attachment and entry of the virus into host cells, antibody directed against these glycoproteins should interfere with infection. To test this hypothesis we purified glycoprotein B (gB), gC and gD using monoclonal antibodies and used them as experimental vaccines in cattle (BABIUK et al. 1987). In all cases the purified protein induced neutralizing antibody and, more importantly, calves were protected from infection. Expression of the glycoproteins in different expression systems clearly indicated that glycosylation or conformation of the glycoprotein was important for inducing the most effective immune responses (VAN DRUNEN LITTEL-VAN DEN HURK et al. 1993a). Glycoproteins produced in *E. coli* did not induce antibody to the epitopes involved in neutralization and therefore was not effective in preventing virus infection. These studies also demonstrated that immunity induced by subunit vaccines was not life-long. Therefore we postulated that immunization of cattle with a plasmid encoding the glycoproteins should result in immunity to the important epitopes and, if the plasmid persisted, it should induce immunity of long duration.

Although various promoters have been used to drive gene expression in vivo, our initial studies used the Rous sarcoma virus (RSV) enhancer/promoter (COX et al. 1993). This expression cassette proved to be satisfactory, but we anticipate that improvements made in plasmid construction will continue to improve the efficiency of expression in vivo. Since many viral glycoproteins are toxic to cells, as is gD, we constructed a number of plasmids expressing different forms of gD (TIKOO et al. 1990). We were especially concerned that if full length gD killed the transfected cell in vivo, the quantity and duration of protection would not be sufficient to induce long-term immunity. Thus we constructed a plasmid containing the entire gD gene and a second plasmid containing the gD gene without the transmembrane anchor (TIKOO et al. 1993). This latter plasmid would secrete the gene product into the extracellular milieu. The third construct lacked the signal sequence and was designed to help determine the possible role of the muscle cell in antigen presentation. In all instances, when the plasmids were introduced into mice, immunity developed (Babiuk et al. 1995a). However, when the antigen was retained in the cytosol there was a delay in the immune response and the magnitude of the immune response was slightly reduced.

Immunization of cattle with the plasmid encoding the full length gD resulted in the development of antibody which could neutralize virus. In addition, plasmid immunized cattle developed lymphocyte proliferative responses similar in magnitude to those found in cattle immunized with purified gD. Based on the observation that serum neutralizing antibodies and cell mediated immune responses were of the magnitude that was previously shown to be protective in cattle, we challenged the animals with an aerosol of a virulent field strain of virus. All control animals exhibited clinical signs characteristic of infection and shed virus for a period of 8–10 days. In contrast, animals immunized with plasmids encoding gD only exhibited mild clinical signs and shed significantly less virus for a much shorter period of time. These results are interesting, since plasmids were introduced via the intramuscular route. If the plasmids were introduced by the intranasal route, one would expect a more significant mucosal response which might totally eliminate virus infection and shedding.

Most reports to date using intramuscular injection of plasmids suggest a polarized Th1 response as measured by the cytokine profiles of purified lymphocytes (Babiuk et al. 1995a). In mice and cattle, immunization with plasmids encoding gD also suggested that a Th1 titer response was produced. These results are important for two reasons. (1) Cell mediated immunity is considered to be critical for recovery from herpes infections (Babiuk et al. 1995b). (2) In BHV-1 especially, interferon-γ plays an important role in activating macrophages to kill BHV-1 infected cells (Campos et al. 1989). Thus a Th1 response is considered to be critical in this disease.

Protection of cattle from BHV-1 infection following immunization with plasmids was extremely encouraging for a number of reasons. First, the observation that DNA immunization can induce immunity in a large animal indicates that it should be possible to adapt this approach to many species of veterinary importance. Second, this method of vaccination provides an opportunity to elicit immunity of long duration against diseases of all species (Michel et al. 1995). For example, horses need to be immunized two to three times per year to protect them against equine rhinotracheitis virus (equine herpes-1). Trainers are generally reluctant to immunize horses during the racing season. Thus, horses often suffer from respiratory disease. Combining an equine herpesvirus plasmid vaccine with plasmids encoding equine influenza virus should dramatically reduce respiratory distress at race tracks. Thus, combining the plasmids encoding for protective antigens of these different pathogens should be effective in horses. Whether improvements in delivery can be made to achieve long-term immunity with a single injection remains to be determined. DNA immunization of poultry and other species against influenza virus with genes encoding for the hemagglutinin and nuclear protein have already been shown to be effective (Robinson et al. 1993a; Webster et al. 1994; Ulmer et al. 1994b; Donnelly et al. 1995).

Pseudorabies virus of pigs is also a potential target for DNA immunization. Presently, many countries are trying to eradicate pseudorabies virus. DNA immunization can be used as a "marker" vaccine such that differential tests can be developed in parallel. A marker vaccine, which only induces antibody to a single glycoprotein, can allow the development of tests to differentiate infected animals from latent carriers. DNA immunization provides such an opportunity. Similar approaches might be applied to tuberculosis or brucellosis against which protection could be provided by vaccination while still allowing the identification of carriers.

The observation that DNA immunization can be effective with viruses and bacteria provides an excellent opportunity to vaccinate animals against multiple viral and bacterial pathogens simultaneously. Furthermore, in modern livestock practices, immunization can only occur at specific times when animals are handled.

Unfortunately, early in the animal's life, a time when it is amenable to vaccination, high levels of maternal antibody are present which interfere with conventional vaccination. The fact that plasmids may persist and might induce immune responses in the presence of passive antibody provides an excellent opportunity to immunize animals earlier in life (Babiuk et al. 1996; Ertl, personal communication).

In addition to inducing immune responses to infectious diseases, DNA immunization also has the potential for control of ectoparasites. Studies have shown that reducing the level of symbiotic bacteria, which supply nutrients to the arthropod, can be a method to reduce arthropod fertility. An example of such an effect is with the Tsetse fly, in which symbiotic bacteria help digest blood meal components and supply nutrients to the Tsetse fly (Nogge 1978). If it was possible to disrupt this symbiotic relationship then the survival of the Tsetse fly would be dramatically reduced. Immunization of animals against the symbiotic bacteria would help neutralize their biological activity in vivo or neutralize the digestive enzymes produced by these bacteria (Allen 1994). This should result in starvation and death of the parasite even if it is engorged. Thus, the overall effect would be a reduction of Tsetse fly population.

Immunization of rabbits against mid-gut antigens of mosquitoes demonstrated lower survival rates in mosquitoes fed on immunized rabbits versus mosquitoes fed on normal rabbits (Alger et al. 1972). Similar results were reported when mid-gut antigens of ticks were used to immunize cattle (Allen et al. 1979). These studies lead various investigators to speculate that it should be possible to use these concealed antigens as vaccines (Willadsen et al. 1993). Concealed antigens do not normally induce immune responses during ectoparasite infestation since they are rarely exposed to the immune system. However, if the host contains antibodies to these antigens in the blood, when the ectoparasite obtains a blood meal, these antibodies interact with the concealed antigen in the gut and in combination with complement may result in lysis or disruption of gut cell function. This results in death of the ectoparasite or a dramatic reduction in its reproductive capacity. Recently an 86 kDa glycoprotein has been identified in ticks which is part of the concealed antigen repertoire. When this antigen was introduced into cattle who were subsequently exposed to ticks, the vaccinated animals had fewer surviving ticks, and these showed reduced engorgement and dramatic reduction of egg production capacity (WILLADSEN et al. 1988). Similar approaches have been used for controlling the sheep blow fly, in which a PM44 protein from the larval gut was identified as the potential concealed antigen (EAST et al. 1993; WILLADSEN et al. 1993; EISEMANN et al. 1994). Since many of these antigens are glycoproteins, administration of genes encoding these glycoproteins in plasmids should induce very high levels of immune responses to the appropriate epitopes. Although there are no reports in the literature indicating that DNA immunization works with concealed ectoparasite antigens, there is confidence that such an approach should work, especially for those ectoparasites which primarily feed on a single host. Thus, if it was possible to immunize all of the hosts in a specific environment it should dramatically reduce the ectoparasite load. If these ectoparasites also serve as transmission vectors for infectious agents, there should be a double benefit in reducing economic losses not only to the ectoparasite but also to the infectious agent transmitted by the parasite.

The type and magnitude of the immune response are influenced not only by the antigen, but also by the micro-environment where the response is being initiated. This micro-environment is influenced by the constellation of cells that are interacting at that site, their level of stimulation and the cytokines they produced. This is best demonstrated by the observation that administration of antigen to mucosal surfaces induces immunity at mucosal surfaces whereas systemic immunization induced systemic immunity. Thus, introduction of a plasmid into mucosal sites, by a gene gun, should induce higher mucosal immune responses than if the genes are introduced intramuscularly or intradermally at sites other than mucosal surfaces. Furthermore, it appears that introduction of plasmids intramuscularly induces primarily Th1 responses with IgG2A as the major antibody isotype. In contrast, gene gun administration favors a Th2 type response and predominantly an IgG1 isotype response (Robinson, personal communication). Thus, if a specific isotype or level of antibody is critical for a specific pathogen, then the appropriate route of administration needs to be chosen. In addition, it may be desirable to drive the immune response by co-administration of cytokines. Since the action of cytokines is extremely short lived, most cytokines need to be administered multiple times to act as adjuvants (HUGHES et al. 1991, 1992a). The administration of plasmids encoding different cytokines could overcome the need for multiple administration of cytokines or slow release delivery systems. To test this hypothesis we co-administered plasmids encoding gD with granulocyte/macrophage colony-stimulating factor (GM-CSF). Results from these experiments suggest that the magnitude of the immune response can be increased by co-administration of GM-CSF with plasmids encoding gD (BABIUK et al. 1996). Since genes coding for many of the animal cytokines have already been cloned, including some cytokines from poultry, the potential use of cytokines to enhance as well as to focus immune responses appears to be feasible for many species. In addition to enhancing immune responses by co-administration of genes encoding cytokines and pathogens the immune modulation induced by these cytokines may provide early protection to animals while immunity is being developed to the specific pathogen of interest.

4 Delivery

DNA immunization provides a number of advantages regarding delivery which should be adaptable to various species and induction of various types of immune responses depending on the agent and the animal which is being immunized. If mucosal immunity is desired, delivery by a gene gun into the mucosal surfaces may be desirable. Although no published reports are available in which delivery to mucosal surfaces with a gene gun has been achieved, Agricetus (Madison, WI) is investigating delivery of such genes to the mucosal surface of the oral cavity of dogs (Haynes, personal communication). This is a relatively easy procedure with minimum discomfort to the animal. Delivery of genes via a gene gun involves coating gold particles with the plasmid of interest. A major advantage of using gold particles to deliver the gene is that elemental gold is inert, the DNA appears to be protected from degradation and is administered directly into the cell. Results to date indicate that immune responses are induced with submicrograms quantities of DNA. This is at least an order of magnitude less than if direct injection is used. Other needle-less injection devices, such as those used for insulin injection in humans, do not use gold particles. In these devices the gene is administered in saline or other liquid vehicles or excipients (Davis et al. 1993). All of these devices need to administer the gene to areas of the body that are devoid of significant body hair and have relatively thin stratum corneum and epithelial layers. In ruminants and other mammals such areas include the oral or nasal mucosa or caudle fold. Another potential site for injection is the inner surface of the ears. The ear appears to be especially attractive since reports have indicated that immunity develops quickly after introduction of genes into the ear (Johnston, personal communication). A further advantage of injection into the ear is that it may induce mucosal immunity due to the shared lymphatic drainage with the nasal mucosal site (Gao et al. 1995). An additional advantage of using the ear as a site of immunization is ease of administration. It is very difficult to restrain large animals to administer vaccines intranasally. This was very evident following the introduction of live intranasal vaccines to cattle. Following administration of the vaccine to a few hundred animals, operator fatigue sets in and the vaccines are not administered properly. Similar concerns will arise with genetic immunization. Although previously we indicated needle injection as being less acceptable, due to broken needles and injection site reactions; subcutaneous injection, even by needles, into the ear would dramatically reduce these problems since the ear is of minimal value to the carcass.

Presently, direct injection into all species requires significantly larger quantities of DNA to induce an immune response than if administered by a gene gun. The reason for this may be degradation of the plasmids before they are taken up by the cells in vivo and transported to the nucleus. To improve the transfection efficiency various groups are developing novel cytofectins or polymers which interact with the DNA to protect it and assist in DNA transport to the nucleus (Felgner 1995; Haensler et al. 1993; Longley et al. 1995). These novel cytofectins should dramatically reduce the quantity of DNA required for injection thereby making the procedure extremely effective and economical, possibly approaching that of the gene gun.

Although developments in the area of cytofectins are progressing rapidly it may be possible to combine two technologies, genetic immunization and amplicons to further improve delivery and immunity (Ho 1994). Using amplicons, genes can be amplified, packaged and introduced into host cells. Since herpes viruses can package approximately 150 kb of DNA within the capsids, multiple foreign genes could be administered with a single vaccination. The viral capsids would protect the DNA extracellularly, allow transport to the nucleus and enhance transfection efficiency dramatically. Delivery of plasmids into mammalian cells by bacterial vectors may also be possible. The economics of producing such vaccines would be significantly better than those of present vaccines since a single amplicon or bacterial vector could induce immunity to a large number of pathogens simultaneously. Furthermore, both of these delivery systems could be used to deliver genes to mucosal surfaces. Techniques for lyophilizing and preserving modified live viruses and bacteria are already in place, thus the downstream processing and vaccine formulation steps for amplicons would not need to be modified dramatically compared to those presently being used by the vaccine manufacturers.

Amplicons would not only be effective delivery vehicles for mammalian species, but would also fit into the in ovo injection system being used by Embrix. In this case the amplicons could be based on Marek's disease which act as the vehicles to introduce genes into the embryo in ovo. This would overcome the impediment of direct DNA injection into embryos in ovo in which introduction of the DNA would need to be administered into the embryo. Presently this appears to be a very difficult if not an impossible task. Using the amplicon it would be possible to deliver the DNA into the embryo and more importantly deliver multiple genes to ensure that the chick is immunized to all of the desired pathogens even prior to hatching.

5 Regulatory Issues

As with any new technology the regulatory agencies and society must ensure both the short- and long-term safety of products licensed for use. Although regulatory agencies evaluate each new vaccine on a case by case basis, some generic information will be important in providing a level of comfort for regulating all nucleic acid vaccines. The first major concern raised is whether the DNA used in the vaccine can integrate into the host genome and subsequently produce deleterious affects to the host. Integration may result in insertional mutagenesis and activation of oncogenes, inactivation of tumor suppressive genes or chromosomal alterations and rearrangements. Since any one of these events could lead to altered cell growth and tumors this would be unacceptable for any vaccine. In the case of integration, numerous attempts have been made at demonstrating whether integration occurs or not. To date, evidence suggests that integration is an extremely rare event, if it occurs at all, with the presently used plasmids. Recent progress in developing new plasmids with minimal or no homology to host sequences as well as utilizing minimal sequences required for in vivo function should further reduce the chance of integration. Thus, it appears that integration and activation of oncogenes by plasmid immunization will be extremely remote and therefore should not be a major concern for licensing DNA vaccines for animals.

The quantity of DNA that remains in the host following genetic immunization is very low and the risk of transfer from the immunized animals to humans consuming the meat is even lower. Therefore, the risk to humans consuming meat from DNA immunized animals should not be an issue. However, to investigate whether tissue containing plasmid DNA may be transferred to individuals consuming the meat, calves were injected with plasmid DNA encoding for bovine herpesvirus-1 gD and 3 weeks later the injection site was excised. This tissue was then fed to rats and several organs and tissues were tested for the presence of gD specific sequences using the polymerase chain reaction. In no instance could we detect any evidence of gD sequences in the animals that consumed tissue injected with plasmid DNA (unpublished results). These results provide a level of confidence that humans consuming meat from animals immunized with plasmids would not be at risk. This appears very logical since a few plasmids that may persist in the tissue of the immunized animals would further be degraded following slaughter, processing and digestion.

Previously, we stated that reactogenicity of vaccine and adjuvant residues was of concern for conventional vaccines. Similar concerns must be addressed with genetic vaccines. Since many of the genetic vaccines being tested to date are either administered without any carriers (intramuscular injection) or by biolistic bombardment with inert gold particles there does not appear to be any local reactogenicity with any of these vaccines reported to date. This is very interesting, especially since the plasmid appears to persist, at least in some instances, with continuous stimulation of the immune system. Investigations to date with numerous plasmids do not show any granuloma formation and only minimal leukocyte infiltration. This is critical for food-producing animals, as the quality of the food is of major concern.

Since there was no evidence of systemic toxicity even when DNA concentrations as high as 1 mg were injected, immunization of pregnant animals should not result in any adverse side affects. Unfortunately, no reports have yet been published in which large doses of plasmid DNA were injected into pregnant animals. An effect of plasmid immunization on reproduction must be performed to determine whether there are any adverse affects or toxicity to the fetus and whether naked DNA can be transferred to the fetus.

Early in the discussion regarding DNA immunization, the concern was raised whether immunization with DNA, which persists, would induce anti-DNA antibodies. It is becoming accepted that it is extremely difficult to induce anti-DNA antibodies, especially if the DNA is not injected with strong adjuvants. Since food-producing animals generally do not live long between immunization and slaughter this concern appears to be minimal. We anticipate that DNA vaccines for veterinary species will be licensed before human vaccines, and the experience with large numbers of veterinary species will provide a level of comfort that anti-DNA antibodies do not develop. This will provide a level of confidence for application of this technology to human vaccination in the future.

The continued expression of an antigen in vivo may induce unexpected consequences such as tolerance or exhaustion of the immune response or, alternatively, autoimmunity or hyperimmunity. Experiments conducted to date suggest that this does not occur, but once again, extensive experience in veterinary species should provide a level of confidence for use of this technology in long-lived species such as humans, in which reversal of adverse reactions would be difficult.

Scale-up, production and purification schemes for DNA vaccines will need to meet the same quality control parameters that other licensed vaccines do today. Thus the product will need to be consistent from batch to batch, both biochemically and genetically. Obviously, impurities in the final product will need to be negotiated with the regulatory agencies; however, certain acceptable parameters are already well established. For example, removal of bacterial by-products, such as endotoxin, is a common practice in vaccine development. In contrast, subunit vaccines require the removal of DNA from the vaccine. Obviously, for DNA vaccines this requirement will need to be modified.

In summary, even though a number of early concerns were raised regarding the potential detrimental effects following DNA immunization, these have not been borne out. Based on these findings, it appears that very few regulatory issues will be hindrances to the licensing of this new generation of vaccines for veterinary species. Furthermore, the experience we gain in veterinary species should have a significant impact on raising the level of comfort for applying this technology to human medicine.

6 Epilogue

The advances in the development of new vaccines for improving animal health have been expedited by the application of biotechnological techniques and our understanding of the host responses to foreign antigens. In parallel with the applications of molecular biology there have been advances in immunology which are further assisting us in our understanding of the host response to immunizations and how these responses can be enhanced. Advances in protein engineering, epitope mapping and understanding the roles of T and B cell epitopes as well as the requirements for specific conformational constraints in inducing immune responses are allowing the development of more efficacious vaccines which stimulate the appropriate immune response. One of the most recent developments in this area is the ability to induce immune responses with plasmids encoding specific protective proteins. By introducing the gene directly into the animal many of the downstream processing steps required for the development of vaccines, adjuvants and formulation are completely eliminated. This makes it extremely attractive for immunizing animals against various diseases and ectoparasites simultaneously. Our ability to re-engineer genes to reduce secondary structure and improve codon bias combined with developments in the area of promoters and delivery vehicles to increase the level of expression in vivo have the potential to dramatically reduce the quantity of DNA required for induction of an effective immune response. Thus, if an immune response can be induced with doses of 1 μg of DNA, animals could be immunized with 10–20 different plasmids simultaneously, provided that no interference occurs between such genes. Clearly, doses as high as 1000 μg do not cause any adverse reactions in animals. A further advantage of plasmid immunization is that each of the individual plasmids could be manufactured and then blended into different combinations of vaccine depending on the management conditions and areas of the globe where the vaccine is being used.

Although the exact mechanism whereby DNA uptake occurs following intramuscular or intradermal injection and which cells actually present antigen are not fully understood, most studies to date suggest that such administration favors a Th1 response. Thus, in parasitic infections and in some bacterial infections, such as mycobacterium, in which a Th1 response is critical, injection may be the most attractive way of vaccination. In contrast, in those cases in which a Th2 response is favored, administration by the gene gun may be more appropriate. Thus, it is envisaged that DNA immunization will allow us to stimulate the type of response which is desired. Furthermore, using plasmids encoding different cytokines it should not only be possible to enhance immune responses, but potentially to tailor the immune response to a Th1 or Th2 type. Finally, since passive immunization may not interfere with development of immune responses following genetic immunization, this new technology should allow genetic immunization to be used under a variety of management conditions including immunization of very young animals. Based on the exciting results observed during the last few years using genetic immunization, we are confident that the vaccines of the future will be much more efficacious and safer than presently licensed vaccines.

References

Alger NE, Cabrera EC (1972) Increase of death rate of Anopheles stephensi fed on rabbits immunized with mosquito antigen. J Econ Entomol 65:165–168

Allen JR (1994) Host resistance to ectoparasites. Rev Sci Tech Off Int Epiz 13(4):1287–1303

Allen JR, Humphreys SJ (1979) Immunization of guinea pigs and cattle against ticks. Nature 280: 491–493

Atanasiu P (1962) Production de tumeurs chez le hamster par inoculation d'acide desoxyribonucleique extrait de cultures de tissues infectees par le virus de polyome. Acad Sci 254:4228–4230

Babiuk LA, Sabara MI, Hudson GR (1984) Rotavirus and coronavirus infections in animals. Prog Vet Microbiol: 80–120

Babiuk LA, L'Italien J, van Drunen Littel-van den Hurk S, Zamb T, Lawman MJP, Hughes G, Gifford GA (1987) Protection of cattle from bovine herpesvirus type 1 (BHV-1) infection by immunization with individual viral glycoproteins. Virol 159:57–66

Babiuk LA, Campos M (1992) Respiratory vaccines for farm animals. In: Peters AR (ed) Vaccines for applications. Butterworth, UK, pp 83–115

Babiuk LA, Lewis PJ, Cox G, van Drunen Littel-van den Hurk S, Baca-Estrada M, Tikoo SK (1995a) DNA immunization with bovine herpesvirus-1 genes. In: Liu M (ed) DNA vaccines: a new era in vaccinology. New York Academy of Sciences, NY, pp 50–56
Babiuk LA, van Drunen Littel-van den Hurk S, Tikoo SK (1995b) Immunology of bovine herpesvirus infections. Vet Microbiol 772:47–63
Babiuk LA, Lewis PJ, Suradhat S, van Drunen Littel-van den Hurk S, Baca-Estrada M, Tikoo SK, Yoo D (1996) Polynucleotide immunization: a novel approach to vaccination. Vaccines 96, Cold Spring Harbor, Cold Spring Harbor, NY, pp 33–38
Campos M, Bielefeldt Ohmann H, Hutchings D, Rapin N, Babiuk LA, Lawman MJP (1989) Role of interferon gamma in inducing cytotoxicity of peripheral blood mononuclear leukocytes to bovine herpesvirus type 1 ~(BHV-1)-infected cells. Cell Immunol 120:259–269
Cox GJM, Zamb TJ, Babiuk LA (1993) Bovine herpesvirus 1: Immune responses in mice and cattle injected with plasmid DNA. J Virol 67:5664–5667
Davis HL, Michel ML, Whalen RG (1993) DNA-based immunization induces continuous secretion of hepatitis B surface antigen and high levels of circulating antibody. Hum Mol Gen 2:1847–1851
Donnelly JJ, Friedman A, Martinez D, Montgomery DL, Shiver JW, Motzel SL, Ulmer JB, Liu MA (1995) Preclinical efficacy of a prototype DNA vaccines: Enhanced protection against antigenic drift in influenza virus. Nature Medicine 1(6):583–587
Donnelly JJ, Ulmer JB, Liu MA (1993) Immunization with polynucleotides: a novel approach to vaccination. The Immunologist 2:20–26
East L, Kerlin RL, Altmann K, Watson DL (1993) Adjuvants for new veterinary vaccines. Prog Vaccin 4:1–28
Eisemann CH, Ginnington KC (1994) The peritrophic membrane: its formation, structure, chemical composition and permeability in relation to vaccination against ectoparasitic arthropods. Int J Parasitol 24:15–26
Felgner PL (1995) Cationic lipids for direct in vivo gene delivery. In: Liu M (ed) DNA vaccines: a new era in vaccinology. New York Academy of Sciences, NY, 772:126–139
Fynan EF, Webster RG, Fuller DH, Haynes JR, Santoro JS, Robinson HL (1993) DNA vaccines: protective immunizations by parenteral, mucosal and gene-gun inoculations. Proc Natl Acad Sci USA 90:11478–11482
Gao Y, Daley MJ, Splitter GA (1995) BHV-1 glycoprotein 1 and recombinant interleukin 1B efficiently elicit mucosal IgA response. Vaccine 13:871
Haensler J, Szoka (Jr) FC (1993) Polyamidoanine cascade polymers mediate efficient transfection of cells in culture. Biconjug-Chem 4(5):372–379
Harland RJ, Potter AA, van Drunen Littel-van den Hurk S, Van Donkersgoed J, Parker MD, Zamb TJ, Janzen ED (1992) Ale effect of subunit or modified live bovine herpesvirus-1 vaccines on the efficacy of a recombinant Pasteurella haemolytica vaccine for the prevention of respiratory disease in feedlot calves. Can Vet J 33:734–741
Ho DY (1994) Amplicon-based herpes simplex virus vectors. Methods Cell Biol 43:191–210
Hoffmann SL, Sedegah M, Hedstrom RC (1994) Protection against malaria by immunization with a plasmodium yoelei circums parazoite protein nucleic acid vaccine. Vaccine 12:1529–1533
Hughes HPA, Babiuk LA (1992a) The adjuvant potential of cytokines. Biotech Therap 3:101–117
Hughes HPA, Campos M, Godson DL, van Drunen Littel-van den Hurk S, McDougall L, Rapin N, Zamb T, Babiuk LA (1991) Immunopotentiation of bovine herpesvirus subunit vaccination by interleukin-2. Immunol 74:461–466
Hughes HPA, Campos M, Potter AA, Babiuk LA ~(1992b) Molecular chimerization of Pasteurella haemolytica leukotoxin to interleukin-2: effects on cytokine and antigen function. Infection and Immunity 60:565–570
Inokuma H, Kerlin RL, Kemp DH, Willadsen P (1993) Effects of cattle tick (Boophilus microplus) infestation on the bovine immune system. Vet Parasitol 47:107–118
Kerlin RL, East U (1992) Potent immununosuppression by secretory/excretory products of larvae of the sheep blowfly Lucilia cuprina. Parasite Immunol 14:595–604
Kit S, Qavi H, Gaines JD, Billinglsey P (1985) Thymidine kinase negative bovine herpesvirus type 1 mutant is stable and highly attenuated in calves. Arch Virol 86:53–83
Kit S, Sheppard M, Ichimura H, Kit M (1987) Second generation pseudorabies virus vaccine with deletions in thymidine kinase and glycoprotein genes. Am J Vet Res 48:780–793
Longley C, Axelrod H, Midha S, Kakarla R, Kogan NA, Sofia M, Babu S, Wierichs L, Walker S (1995) Conjugates of glycosylated steroids and polyamines as novel non-viral gene delivery systems. In: Liu M (ed) DNA vaccines: a new era in vaccinology. New York Academy of Sciences, NY, 772:268–270

Lowrie DB, Tascon RE, Colston MJ, Silva CL (1994) Towards a DNA vaccine against tuberculosis. Vaccine 356:152–154

Michel MIL, Davis HL, Schleef M, Manicin M, Tiollais P, Whalen RG (1995) DNA-mediated immunization to the hepatitis B surface antigen in mice: aspects of the humoral response mimic hepatitis B viral infection in humans. Proc Natl Acad Sci 92:5307–5311

Morein B, Lovgren K, Hoglund S, Sundquist B (1987) The ISCOM. An immunostimulating complex. Immunol Today 8:333–338

Nick S, Klaws J, Friebel K, Birr C, Hunsmann G, Bayer H (1990) Virus neutralizing and enhancing epitopes characterized by synthetic oligopeptides derived from feline leukemia virus glycoprotein sequence. J Gen Virol 71:77–83

Nogge G (1978) Aposymbiotic tsetse flies, Glossina morsitans morsitans, obtained by feeding on rabbits immunized specifically with symbionts. J Insect Physiol 24:299–304

Olsen CW, Corapi WV, Ngichake CK, Baines JD, Scott FW (1992) Monoclonal antibodies to the spike protein of feline infectious peritonitis virus mediate antibody dependent enhancement of infection of feline macrophages. J Virol 66:956–965

Opdebeeck JP (1994) Vaccines against blood-sucking arthropods. Vet Parasitol 54:205–222

Pastoret PP, Brochier B, Languet B, Thomas I, Paguot A, Bauduin B, Kieny MP, Lecocq JP, DeBruyn J, Costy F, Antonine H (1988) First field trial of fox vaccination against rabies using a vaccinia-rabies recombinant virus. Vet Rec 123:481–483

Pedersen NC, Johnson L, Bird C, Theilen GH (1986) Possible immunoenhancement of persistent viremia by feline leukemia virus envelope glycoprotein vaccines in challenge-exposure situations where whole inactivated vaccines were protective. Vet Immuno and Immunopathol 11:123–148

Potter AA, Harland RJ (1990) Development of a recombinant subunit vaccine for Pasteurella haemolytica. American Society of Microbiology Biotechnology Conference, June 1990, Chicago, IL

Robinson HL, Hunt LA, Webster RG (1993a) Protection against a lethal influenza virus challenge by immunization with a haemagglutinin-expressing plasmid. DNA Vaccine 11(9):957–960

Robinson HL, Hunt LA, Webster RG (1993b) Protection against a lethal influenza challenge by immunization with a hemagglutinin expressing plasmid. DNA Vaccine 11:957–960

Tang DC, DeVit M, Johnston SA (1992) Genetic immunization is a simple method for eliciting an immune response. Nature 356:152–154

Theodos CM, Titus RG (1993) Salivary gland material from the sandfly Lutzomyia longipalpis has an inhibitory effect on macrophage function in vitro. Parasite Immunol 15(8):481–487

Tikoo SK, Fitzpatrick DR, Babiuk LA, Zamb TJ (1990) Molecular coloning, sequencing, and expression of functional bovine herpesvirus 1 glycoprotein gIV in transfected bovine cells. J Virol 64:5132–5142

Tikoo SK, Zamb TJ, Babiuk LA (1993) Analysis of bovine herpesvirus 1 glycoprotein gIV truncations and deletions expressed by recombinant vaccinia viruses. J Virol 67:2103–2109

Tikoo SK, Campos M, Babiuk LA (1995) Bovine herpesvirus-1 (BHV-l): Biology pathogenesis and control. Adv Virus Res 45:191–223

Ulmer J, Deck RR, DeWitt CM, Friedman A, Donnelly J, Liu M (1994a) Protective immunity by intramuscular injection of low doses of influenza virus DNA vaccines. Vaccine 12:1541–1545

Ulmer JB, Deck RR, DeWitt CM, Friedman A, Donnelly JJ, Liu MA (1994b) Protective immunity by intramuscular injection of low doses of influenza virus DNA vaccines. Vaccine 12:1541–1544

Ulmer JG, Donnelly JJ, Parker SE, Rhodes GH, Felguer PL, Dwarki VJ, Gromkowski SH, Deck RR, DeWitt CM, Friedman A, Hawe LA, Leander KR, Martinez D, Perry HC, Shiver JW, Montgomery DL, Liu MA (1993) Heterologous protection against influenza by injection of DNA encoding in viral protein. Science 259:1745–1749

Valenzuela P, Medina A, Rutter WJ, Ammerer G, Rutter BD (1982) Synthesis and assembly of hepatitis B surface antigen particles in yeast. Nature 298:347–350

van Drunen Littel-van den Hurk S, Parker NW, Massie B, van den Hurk JV, Harland R, Babiuk LA, Zamb TJ (1993a) Protection of cattle from BHV-1 infection by immunization with recombinant glycoprotein gIV. Vaccine 11:25–35

van Drunen Littel-van den Hurk S, Tikoo SK, Liang X, Babiuk LA (1993b) Bovine herpesvirus-1 vaccines. Immunol Cell Biol 71:405–42

Webster RG, Fynan EF, Santoro JC, Robinson H (1994) Protection of ferret against influenza challenge with a DNA vaccine to the haemagglutinin. Vaccine 12(16):1495–1498

Weiss WR, Berzovsky JA, Houghten R, Sedegah M, Hollingdale M, Hoffman SL (1992) A T cell clone directed at the circumosporozoite protein which protects mice against both P yoelli and P berghei. J Immunol 149:2103–2109

Whitton IL, Sheng N, Oldstone M, McKee T (1993) A "strong of beads" vaccines, comprising linked minigenes, confers protection from lethal virus challenge. J Virol 67:348–352

Wikel SK, Ramachandra RN, Bergman DK (1994) Tick-induced modulation of the host immune response. Int J Parasitol 24:59–66

Willadsen P, Eisemann CH, Tellam RL (1993) "Concealed antigens" expanding the range of immunological targets. Parasitol Today 9:132–135

Willadsen P, McKenna RV, Riding GA (1988) Isolation from the cattle tick Boophilus microplus of antigenic material capable of eliciting a protective immunological response in the bovine host. Int J Parasitol 18:183–189

Williams RS, Johnston SA, Reidy M, Devit MJ, McElligott SC, Sanford JC (1991) Introduction of foreign genes into tissues of living mice by DNA-coated microprojectiles. Proc Natl Acad Sci USA 88:2726–2730

Wolff JA, Malone RW, Williams P, Chong W, Acsadi G, Jani A, Felgner PL (1990) Direct gene transfer into mouse muscle cells in vivo. Science 247:1465–1468

Xiang ZQ, Spitalnik S, Tran M, Wunner WH, Cheng J, Ertle HC (1994) Vaccination with a plasmid vector carrying the rabies virus glycoprotein gene induces protective immunity against rabies virus. Virology 199:132–140

Xu D, Liew FY (1994) Genetic vaccination against leishmaniasis. Vaccine 12:1534–1536

Yilma T, Hsu D, Jones L, Owens S (1988) Protection of cattle against rinderpest and vaccinia virus recombinants expressing the HA and F genes. Science 242:1058–1061

Zijl M, Wensvoort G, de Kluyver E, Hulst M, van den Gulden H, Gielkens A, Bems A, Moorman R (1991) Live attenuated pseudorabies virus expressing envelope glycoprotein El of hog cholera virus protects swine against both pseudorabies and hog cholera. J Virol 65:2761–2765

Nucleic Acid-Based Vaccines as an Approach to Immunization Against Human Immunodeficiency Virus Type-1

M.L. Bagarazzi, J.D. Boyer, V. Ayyavoo, and D.B. Weiner

1 Introduction

The spread of HIV-1 has resulted in significant morbidity and mortality worldwide. Traditional therapeutic approaches have included chemotherapy directed at several different viral components either alone or in combination. Studies evaluating combinations of two or three different anti-retroviral agents have recently generated optimism regarding the ability to control viral replication in vivo. It is too

Department of Pediatrics, Allegheny University of the Health Sciences and Department of Pathology and Laboratory Medicine, University of Pennsylvania School of Medicine, Philadelphia, PA 19104, USA early to tell whether this approach is capable of controlling established infection without additional measures including immunotherapy. In any event, the majority of the worldwide at-risk population will not have easy access to these expensive therapeutic regimens. If achievable, protective vaccination represents the best solution to the worldwide problem of infection with HIV-1 and any strategy designed to control the epidemic is likely to rely heavily on mass immunization campaigns (MUSGROVE 1993). Unfortunately, the effort to design effective vaccines for HIV-1 has encountered a number of significant obstacles. In a 1993 poll by the journal Science (COHEN 1993) 150 top AIDS researchers agreed that the most significant obstacle has been the lack of consistent correlates of protection from natural or experimental infection. It is therefore not surprising that a wide variety of therapeutic modalities have been proposed in pursuit of effective therapy for HIV-1 (Table 1).

Novel as well as traditional immunization strategies are under investigation for the prevention and therapy of HIV-1 infection. In addition, so-called gene therapy,

Table 1. Therapeutic approaches to HIV-1 infection

Traditional chemotherapy
Reverse transcriptase nucleoside analogues
Reverse transcriptase non-nucleoside analogues
Protease inhibitors
Gene therapy
Nucleic acid vaccines (*e.g.* plasmid DNA, RNA)
Interference with viral replication
Insertion of nonpathogenic HIV-1 genes using numerous proposed vectors
Use of decoy RNA (*e.g.* TAR, RRE)
Ribozymes
Antisense RNA for halting HIV-1 mRNA translation to protein
Trans-dominant proteins to compete with viral proteins for functional binding sites
Tat-driven cellular toxins
"HIV homicide" – (e.g. genetic construct coding for Vpr combined with toxic protein)
Induction of antiviral immunity
Intracellular antibodies to HIV proteins
Immunization strategies
Live attenuated with multiple gene deletions
Whole inactivated
Recombinant protein subunit/peptides (e.g. gp120, gp160), lipopeptides
Hepatitis B surface antigen as carrier protein for HIV proteins or peptides
Novel vectors
Adenovirus
Canarypox with or without recombinant protein/peptide in prime-boost strategy
Influenza virus
Listeria monocytogenes
Poliovirus minireplicons
Salmonella typhi
Streptococcus gordonii
Vaccinia
Vector transduced autologous fibroblasts
Nucleic acid
Plasmids with and without facilitators
Gold bombardment (gene gun)

Table 2. Immune responses induced by different vaccine approaches

	Helper T cell					
Vaccine approach	Th1	Th2	CTL	Antibody	Strengths	Weaknesses
Live attenuated	?	?	+	+	Cellular immunity Humoral immunity	Safety in immuno-compromised
Whole inactivated		+	–	+	Ease of preparation Cost	Limited immunity
Virus-like particles		+	–/+	+		Limited immunity Difficult preparation
Protein subunit/peptides		+	–/+	+	Humoral immunity	Poor cellular immunity Difficult preparation Cost
Plasmid DNA	+		+	+	Cellular immunity Humoral immunity Ease of preparation Cost	Limited experience

designed to interrupt viral replication via the manipulation of DNA, mRNA or other cellular components, is currently under development. One less traditional approach which may be considered a fusion of immunotherapy and gene therapy is vaccination with nucleic acid constructs. In theory, direct genetic immunization mimics aspects of live attenuated vaccines in that both humoral and cellular immune responses are induced while avoiding the risks of infection with live virus (see Table 2). The theory has now been verified by a number of investigators studying several different viruses in several different animal models. A summary of these efforts as they pertain to vaccines for HIV-1 are contained herein.

2 Human Immunodeficiency Virus Type-1 Life Cycle and Genome

HIV-1 is an RNA retrovirus belonging to the lentivirus family. The HIV life cycle begins with viral penetration into the lipid bilayer via interaction of viral gp120 envelope glycoprotein and cellular CD_4^+ receptor as well as the relevant second receptor. Briefly, the envelope protein undergoes a series of conformational changes after binding to CD_4^+. Conformational changes are induced by CD_4^+ but it appears that further interaction is required to induce the necessary changes in conformation for the virus to gain entry into its host cells. This interaction has recently been elucidated to require the presence of coreceptors for gp120. Briefly, two classes (four separate chemokines) of soluble CD_8^+ derived inhibitor(s) of HIV-1 infection originally sought by Levy and coworkers (Mackewicz et al. 1994) were discovered to have a role in HIV-1 biology. The β-chemokines (MIP-1α, MIP-1β and RANTES) appear to fit this role with regards to primary HIV-1 isolates (so-called macrophage-tropic virus) (Cocchi et al. 1995). Berger's group also discovered (Feng et al. 1996) a T cell tropic HIV-1 gp120 coreceptor originally dubbed fusin and now known as LESTR (leukocyte-expressed seven-transmembrane-domain receptor). This was followed by the identification of the β-chemokine coreceptor CC R5 found on CD_4^+ cells infected with primary strains of HIV-1 (Bleul et al. 1996; Oberlin et al. 1996). Most recently, two independent teams (Bleul et al. 1996; Oberlin et al. 1996) simultaneously reported that the natural ligand of LESTR is a previously known chemokine, stromal cell-derived factor-1 (SDF-1), which efficiently inhibits infection of CD_4^+ cells by T cell tropic virus. In the context of vaccines, immune responses which target regions of the envelope required for interaction with the chemokine receptors could block viral entry. Additionally, vaccines which induce immune responses resulting in production of relevant chemokines could similarly inhibit viral entry.

The outer lipid layer of HIV acquired when the viral particle buds from the cellular plasma membrane surrounds a protein core that contains the RNA genome (Greene 1991). Once the viral lipid envelope fuses with the cellular membrane the viral core enters the cell. This is followed by the synthesis of a double stranded DNA version of the HIV genome by the viral DNA polymerase or reverse transcriptase (RT). This so-called DNA provirus is then translocated to the nucleus as part of a protein-DNA pre-integration complex and is then integrated into the host cell genome with the help of the viral integrase (Int) enzyme. Once this occurs the provirus is replicated along with the host DNA each time the cell divides.

In the course of reverse transcription integration of the HIV RNA genome two identical long terminal repeats (LTRs) containing enhancer and promoter sequences are generated at each end of the provirus. As opposed to other retroviruses, the HIV-1 genome is organized into three major genes and four accessory and two regulatory genes (Cullen et al. 1990). The canonical retroviral genes are *gag*, which codes for the core protein; *pol*, which codes for the enzymatic proteins RT, Int, and protease (Pro); and *env* which codes for the outer viral envelope proteins. The *gag* gene start site is at the 5′ end of the genome and codes for a precursor polyprotein (Pr55) which is cut to produce four smaller internal structural proteins by the viral enzyme Pro. Pro actually is derived from a second longer polyprotein Pr160 which is produced from the *gag-pol* RNA. RT and Int are also derived from Pr160. *Env* codes for a large glycoprotein, gp160, which is glycosylated and later cut into two smaller molecules, gp120 and gp41 by a host protease as viral particles bud from the cell membrane.

Transcription of the integrated provirus begins when enhancer and promoter sequences of the LTR become active. Fully spliced viral RNAs leave the nucleus and move to the cytoplasm in the early stage. These early transcripts code for viral proteins with regulatory effects on HIV gene expression. In late stage gene expression, partially spliced RNAs are produced that code for viral structural proteins and enzymes as well as full length RNA for the viral genome. The enzymes coded by *pol* and proteins coded by *gag* assemble around the genomic RNA to form the inner viral core. Viral cores acquire the *env* gp120, gp41 and cellular lipid bilayer as they bud from the cell surface. Most HIV are nonlytic although as a lentivirus they may exhibit cytopathic effects on infected cells.

The accessory (*vif*, *vpr*, *vpu* and *nef*) and regulatory (*tat* and *rev*) genes may provide important targets for the therapy of HIV in general and specifically for the design of genetic vaccines for HIV-1. The first to be studied in detail was *tat*, which codes for the protein of the same name, Tat. It has been shown to increase gene expression in infected cells several hundredfold. This is mediated via *trans*-activation by transcript enlargement through an RNA-derived target within the LTR termed the TAR (*trans*-acting response) element (ROSEN 1991). Since Tat increases HIV gene expression dramatically it is an obvious target for therapeutic intervention. One could argue whether *tat* should be included in any genetic vaccines designed to control HIV infection. While it may increase overall expression of the other desired immunogenic sequences of the vaccine, one could also argue that expression of wild-type HIV could be theoretically increased in already infected individuals by providing exogenous *tat*. Accordingly, a nonfunctional yet immunogenic Tat immunogens would be preferred vaccine targets due to its essential role in the viral life cycle and its *trans*-activating activity.

Rev protein is involved in the transition from the production of early, highly spliced RNA to the later production of less extensively spliced RNA encoding structural proteins. Due to this critical function, Rev clearly plays a crucial role in HIV infection. Rev appears to increase the release of unspliced structural RNAs from the nucleus through its interaction with the so-called Rev response element (RRE). The mechanism appears to involve displacing splicing factors from the RNA which otherwise prevent RNA transport from the nucleus to the cytoplasm (MALIM et al. 1989). Rev is a critical component in the production of the structural proteins of HIV-1. The immune response against Rev could target this essential gene sequence and thus interfere with the virus life cycle.

The accessory genes were originally thought to be nonessential; however, their high degree of conservation and association with increased viral protein expression in in vitro assays disputes this assertion. For instance, the *nef* (negative factor) gene had never been shown to be critical for viral infection of cell lines in vitro (Kim et al. 1989). However, experiments performed with the related simian immunodeficiency virus (SIV) found that rhesus macaques infected with virus containing a deletion in *nef* had dramatically lower levels of viral replication and, perhaps more importantly, no evidence of the pathogenesis of SIV infection was observed throughout follow-up (KESTLER et al. 1991). This was in sharp contrast to animals infected with either wild-type *nef* or virus containing a stop codon within *nef* which produced significantly higher levels of virus; several animals (5/12) went on to develop significant immunodeficiency and ultimately expired. Nef also appears to provide an important target for cytotoxic T lymphocytes (CTLs). Based on these observations, Nef should be considered as a target antigen in the design of genetic vaccines for HIV-1.

The final three regulatory genes of HIV, *vif* (virion infectivity factor), *vpr* (viral protein r) and *vpu* (viral protein u) are beginning to receive closer scrutiny. Virus deficient in *vif* has been shown to be about 100-fold less infectious than wild type for cells growing in tissue culture. *vif* was also found to be present at the late stage of replication (FISHER et al. 1987). Experiments using *vif* mutants revealed that its expression was required for infectivity of peripheral blood mononuclear cells (PBMCs) (FAN et al. 1992; GABUZDA et al. 1992) and macrophages (VON SCHWEDLER et al. 1993). As a result, one might include mutant *vif* sequences in any genetic vaccines for HIV-1 or take advantage of its product as a target amenable to immune attack.

The role of Vpr protein in HIV-1 pathogenicity and infection is less clear but certain concepts have emerged. Mutants of *vpr* exhibit markedly diminished replication compared to wild type, particularly in macrophages (CONNOR et al. 1995). Vpr protein may also play a role in cell death and proliferation as cells infected with functional Vpr are unable to sustain infection and perish (LEVY et al. 1995; ROGEL et al. 1995). Vpr interferes with cellular proliferation by interrupting the cell cycle at the transition from G2 to M phase (JOWETT et al. 1995; ROGEL et al. 1995). Vpr has also been shown to interact with putative cytosolic receptor proteins (NARAYAN et al. 1995; REFAELI et al. 1995). These mutants of SIV *vpr* and its duplicated sister gene *vpx* are similar to *nef* deletions in that they appear to be nonpathogenic. These observations would lead one to speculate that *vpr* and its product play an important if not critical role in the life cycle of HIV and possibly in disease pathogenesis. It would appear that *vpr* could be an important component in candidate vaccines. However, once again, its function should be ablated while conserving its immunogenicity.

vpu is ultimately translated into a late viral product which is found exclusively in HIV-1 isolates but not in the other primate lentiviruses (i.e., SIV, HIV-2). It has been found to be nonessential for in vitro replication in PBMCs (Lenburg et al., unpublished observations), although replication increases modestly in cells infected with wild-type *vpu* as compared to mutants (KLIMKAIT et al. 1990). Vpu has also been shown to down-regulate CD_4^+ expression and increase virion release. However, *vpu* is about as well conserved as *vif* and therefore remains a viable immune target which can be exploited by genetic vaccination techniques.

Recent studies of long-term nonprogressors appear to support the importance of the role of accessory genes in in vivo pathogenesis. Nonprogressor cohorts infected with virus containing *nef* deletions, *vif* deletions and, importantly, *vpr* deletions have been reported (MICHAEL et al. 1995; WANG et al. 1996). These data indicate that cellular responses targeting these proteins could directly attenuate viral replication in vivo. This could ultimately result in attenuation of viral phenotypes in vivo even in the absence of viral clearance.

The aforementioned body of work outlining the different functions of HIV genes and their products provides a framework for advancing to the design of candidate genetic vaccines. The information gathered to date on characteristics of HIV which create obstacles for vaccine development are equally important and are described in the section to follow.

Table 3. Major obstacles in HIV-1 vaccine development

Troublesome and often unique biologic and epidemiological viral characteristics
Antigenic diversity including geographic variation
Cell-free and cell-associated infection via mucosal routes
High prevalence within impoverished populations
Establishment of long-term clinical not biological latency
Sequestration in immune-privileged sites (CNS, lymphoid organs) (Fauci 1993)
Difficult recognition of nonspecific acute infection syndrome
Rare evidence for "spontaneous recovery" from infection
Inherent immune suppression
Immunological inaccessibility due to proviral DNA integration into host genome
Major gaps in knowledge of viral pathogenesis and the correlates of protective immune responses
The lack of a satisfactory animal model for infection and disease

Table 4. Major unanswered questions in HIV-1 vaccine development

Are there major cross-reactive epitopes even across viral subgroups which will provide suitable targets for humoral or cellular responses?
How will infection of sequestered sites such as the CNS be prevented and would an otherwise effective therapeutic vaccine have any effect on virus already resident in the CNS?
What are the critical components of an immune response that blocks infection at the mucosal level especially as it relates to cell-associated infection?

3 Obstacles to Vaccine Development

There are several major obstacles to the development of effective vaccines against HIV-1 (Table 3). A great deal has been learned about HIV-1, ranging from extensive knowledge of its genetic makeup to the elucidation of its binding to cellular receptors providing entry into CD_4^+ cells. However, significant gaps remain in our knowledge. Some of the most compelling are listed in Table 4. Perhaps the greatest strides in vaccine development will result by answering even a portion of these questions.

3.1 Inherent Characteristics of Human Immunodeficiency Virus Type-1

Aside from gaps in our knowledge, vaccine development has been hindered by several known characteristics of HIV-1. For instance, the antigenic diversity of HIV-1 alone presents formidable problems for vaccine developers. As an RNA virus whose polymerase (RT) lacks an editing function HIV-1 is naturally subject to a significant amount of sequence divergence over time (Walker 1994). Divergence occurs in an infected individual (intra-individual), or so-called quasispecies (Saag et al. 1988), as well as within the total worldwide pool (inter-individual) of HIV-1 (Fig. 1). The diversity seen among quasispecies may be responsible for the rapid emergence of resistance to anti-retroviral therapy as variants appear to arise in vivo

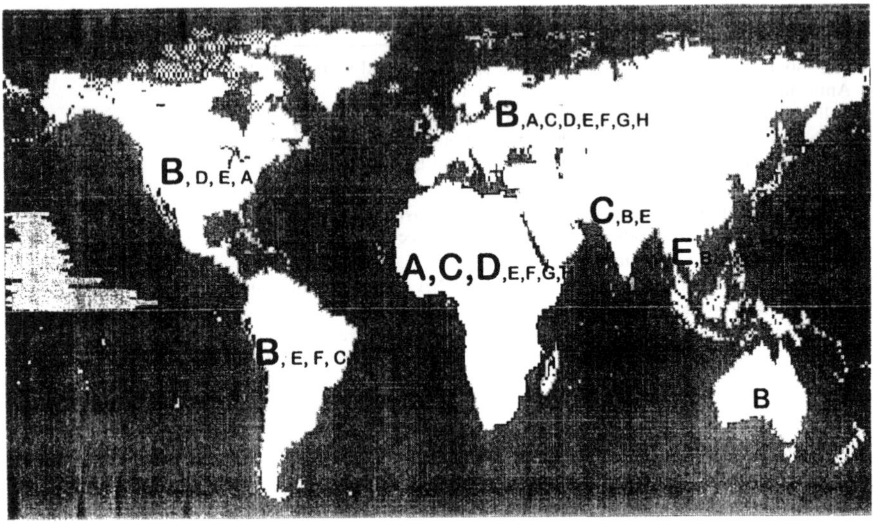

Fig. 1. Geographic distribution of HIV-1 subtypes in the 1990s. (Modified from Birx et al., personal communication)

from a common ancestor virus. The impact that this less extensive intra-individual diversity has on vaccine development suggests that escape from the immune response is not limited to the more dramatic variability observed between viral subtypes.

The inter-individual variability among viral subtypes generally follows geographic lines with sequence variability of envelope proteins measured to be 20%–30% between subtypes (Graham et al. 1995). This diversity is manifested clinically by the less than promising titers of neutralization antibody measured in response to challenge with heterologous strains. One possible bright spot lies in the fact that the antigens which elicit CTL-mediated responses may prove to be less divergent. Internal structural proteins, particularly those with enzymatic function, i.e., RT, protease and integrase, must conserve domains in order to preserve their catalytic function. Consequently, these proteins exhibit less antigenic diversity than envelope proteins but appear to elicit the strongest CTL responses (Nixon et al. 1988). These proteins are processed in the cytoplasm and presented as peptides in association with class I MHC molecules. Functional loss of CTL responsiveness to these conserved proteins has been reported in association with sequence mutations (Phillips et al. 1991) though less frequently than what has been observed with the viral envelope proteins.

The existence of cell-free and even inefficient cell-associated infection via mucosal routes (Anderson et al. 1992) probably mandates the participation of the host's cellular immune apparatus in concert with secretory IgA and perhaps other uncharacterized elements. Knowledge is limited regarding the elements necessary for protection at mucosal sites and the ability of candidate vaccines to elicit them. It is unlikely that antibodies will play a major role in protection from cell-associated infection. Nonetheless, SIV-specific antibodies have been found in rectal secretions of macaques which resisted intrarectal challenge with cell-free SIV after immunization with a whole inactivated virus (CRANAGE et al. 1992). Preliminary studies (PAUZA et al. 1993) in macaques have shown that intrarectal inoculation of SIV_{mac251} produces a lower viral burden than an equivalent intravenous dose. However, another SIV vaccine (ROSENTHAL 1993) demonstrated protection from intravenous but not mucosal challenge. Another potential problem is the infection of Langerhans cells at mucosal surfaces where they may proliferate and disseminate virus through the lymphatics creating systemic infection (HILLEMAN 1992). It would currently appear that the lone area of consensus regarding mucosal immunity is the need for more data, as any vaccine which hopes to be protective in humans will almost surely require a robust response at mucosal sites.

Practical considerations dictate that any vaccine intended for use among developing nations should be both inexpensive and easy to administer. Ease of administration will necessarily require the vaccine protection to be long-lived as well as potent, requiring fewer boosts. Vaccine use in immunotherapy will be limited by the virus's tropism for certain sequestered sites. Vaccine by-products may ultimately have to reach previously inaccessible areas such as the central nervous system (CNS).

The leukocyte tropism and the immune dysfunction ultimately caused by HIV-1 may present the most unique and troublesome barriers to vaccine development. Any attempt to use vaccines as immunotherapy after the establishment of infection could be complicated by viral variability. Even the introduction of non-replicative virus runs the minute yet real risk of creating a revertant virulent strain. Ignoring this risk, it is still difficult to predict how the host will respond to this constant immune stimulation. This is especially important in light of the mounting evidence of viral activity and turnover when other less sensitive indicators would indicate a latent infection. An HIV-infected immune system appears to be quite preoccupied in maintaining the outward appearance of controlled replication according to the data showing extensive viral replication and clearance (HO et al. 1995; WEI et al. 1995). Other potentially troubling features of HIV-1 include evidence supporting the induction of T cell anergy and even apoptosis induced by envelope protein (GROUX et al. 1992; TAYLOR 1992) and direct cellular pathogenesis mediated by the viral accessory genes including *vpr* (CONNOR et al. 1995; LEVY et al. 1993; ROGEL et al. 1995). Similarities between host antigens and regions of the virus also raise the risk of autoimmunity being mediated by some immunization strategies (GOLDING et al. 1988).

3.2 Animal Models

HIV-1 vaccine research has suffered from the use of less than perfect animal models in which infection is not necessarily followed by disease. SIV infects small primates and is closely related to HIV-2 but less so to HIV-1. In spite of this fact, it is still

Table 5. Comparison of different primate models for use in HIV vaccine research

Feature	Chimpanzee/HIV-1	Monkey/SIV and HIV-2	Monkey/SHIV
Commonality with human immune system	+ + +	+ +	+ +
Antigenic cross-reactivity of divergent MHC antigens	+ + +	+ +	+ +
Antigenic cross-reactivity of T cell receptor antigens	+ + +	+ +	+ +
Molecular and biological conservation of cytokines and co-stimulatory molecules	+ + +	+	+
Ability to isolate virus from PBMCs despite the presence of HIV-1 specific neutralizing antibodies and possibly CTLs	+ + +	?	?
HIV-1 isolates derived from infectious human material that is known to be pathogenic in the natural host establishes persistent infection in challenged primate	+ =	–	–
Development of disease after infection	Under certain conditions	+	±

currently one of the models of choice. The majority of HIV-1 strains do not infect these animals yet an immunodeficiency similar to AIDS does occur in macaques infected with SIV_{mac}. The more closely related chimpanzees have not exhibited disease after infection with SIV_{cpz} and until recently HIV-1 did not appear to cause disease in these animals. However, it does appear that the first evidence of disease induced by HIV-1 infection in a few chimpanzees is forthcoming (Novembre et al. 1996). In this regard, infection of chimpanzees with virulent isolates represents a true animal model of HIV-1 infection and pathogenesis (Table 5). However, the prolonged clinical latency period and low number and expense as well as the ethical concerns surrounding experimental use of this primate limits the broad use of this model.

Other potentially useful avenues of investigation include the use of SIV/HIV-1 chimeras which are designed with envelope proteins of HIV-1 while maintaining SIV internal proteins which lead to infection of macaques (Li et al. 1992). Some investigators have also attempted to identify lower primates such as *Macaca nemestrina* which are more susceptible to infection with HIV-1 (Agy et al. 1992). Ultimately, the same poorly understood interactions between virus diversity and narrow host range may be the limiting factor in finding the perfect model. Based on these problems, an ideal candidate vaccine would be capable of conferring protection in a wide variety of challenge systems. While this is a high standard, any vaccine strategy which confers protection in multiple animal models will provide the most confidence possible for subsequent human trials.

3.3 Lack of Correlates of Protective Immunity

3.3.1 Potential Humoral Correlates

The search for strong correlates of protective immunity against HIV-1 is one of the most important and active areas of research in the overall effort to combat HIV-1. Humoral immunity in the form of antibody production provides a significant contribution to protective responses in some successful viral vaccines. Antibodies are especially critical in preventing susceptible cells from becoming infected by blocking virus binding and/or entry to target cells. The presence of high-titer so-called neutralizing antibodies has been correlated with efficacy in a number of viral diseases including hepatitis B and rabies. Consequently, several studies of putative SIV, HIV and S/HIV immunogens have concentrated on assessing this immunological measure.

Strong neutralization activity was found in a study of long-term non-progressors (CAO et al. 1995). Cell-free infection with HIV-1, SIV_{SM} and HIV-2 has been successfully prevented by monoclonal and polyclonal antibody preparations, respectively (EMINI et al. 1992; PUTKONEN et al. 1991). Several HIV-1 vaccines have already been shown to induce neutralizing antibodies, especially those directed at the envelope proteins gp120 and gp41 (GRAHAM et al. 1996; SCHWARTZ et al. 1993). The presence of neutralizing antibody has usually correlated with protection from homologous challenge. The most important epitopes appear to be the hypervariable V3 loop and CD_4^+ binding domain of gp120 as well as sites in gp41. Human vaccine trials to date have shown some cross-neutralization and primate studies have shown limited protection from challenge using laboratory isolates such as LAV/III_B as an immunogen (BARRETT et al. 1991; BERMAN et al. 1990; BRUCK et al. 1993a, b; GIRARD et al. 1991). However, HIV-1 naive vaccine volunteers and seropositive individuals all show higher neutralizing antibody titers to these laboratory isolates. Another study (GIRARD et al. 1995) demonstrated protection from a low dose intrasubtype B heterologous $HIV\text{-}1_{SF2}$ challenge in two chimpanzees vaccinated with gp160 and V3 loop peptides from $HIV\text{-}1_{MN}$ which appeared to correlate with antibody titers to $V3_{MN}$ and neutralizing antibody titers to $HIV\text{-}1_{MN}$ and $HIV\text{-}1_{LAI}$ but not $HIV\text{-}1_{SF2}$. However, neutralization has not always correlated with protection from viral challenge (CHENG-MAYER et al. 1988; KATZENSTEIN et al. 1990). Finally, neutralization of HIV-1 field isolates has not been unambiguously demonstrated to date, perhaps due to divergent laboratory isolates or features of the cells used in the assays. Alternatively, a high frequency of adhesion molecule interactions on activated PBMC targets serving as alternate receptors for viral entry may serve to obscure true viral neutralization by antibody (Zolla-Pazner et al., personal communication).

The importance of antigenic diversity was predicted by showing that neutralizing antibodies developed against the V3 loop of a United States isolate were unable to neutralize African viral isolates (CHEINGSONG-PAPOV et al. 1992). In addition, natural history studies of HIV-1 infection have shown type-specific neutralizing antibody directed at the V3 loop to be a poor correlate of immunity

(Sheppard et al. 1994). Mucosal immunity may also provide a component of successful protection from HIV-1 as mentioned above, as well as the inhibition of syncytium formation and subsequent prevention of cell-to-cell viral spread. None of these measures has clearly been shown to correlate with protection. However, discussants at the Conference on Advances in AIDS Vaccine Development did arrive at the consensus opinion that complex humoral immune responses involving multiple linear and conformational epitopes may be required for efficient and broad neutralization and therefore future trials should attempt to assess this activity (Sheppard et al. 1994).

3.3.2 Potential Cellular Correlates

Cytotoxic T cell (CD_8^+) induced cytolysis has also been shown to correlate with protection in some viral vaccine models including HIV-1. Specific CTL activity has been observed in virtually every case of acute HIV-1 infection studied, with several investigators having shown an association with viral clearance (Borrow et al. 1994; Koup et al. 1994). CTL responses have been observed in response to internal structural proteins as well as HIV-1 envelope (Riviere et al. 1989), with stronger responses noted in response to the former (Walker et al. 1988). Several so-called long-term nonprogressors (Cao et al. 1995; Pantaleo et al. 1995) possess particularly active CD_8^+ T cells which are able to block the capacity of their own CD_4^+ T cells to infect other PBMCs. Since loss of cellular immune function has been shown to be a hallmark of HIV-1 disease progression it would not be surprising to find that the presence of strong cellular responses might correlate with protection. Indeed, a number of the anecdotal reports of presumed successful protection from HIV-1 infection have shown specific CTL activity. Shearer and colleagues (Pinto et al. 1995) have identified a subset (seven of 20) of occupationally exposed but uninfected health care workers with evidence of a transient but HIV-1 specific CTL response. CTLs targeted at specific HIV-1 epitopes were also found in a number of chronically exposed prostitutes in Gambia who continue to resist infection with HIV-1 (Rowland-Jones et al. 1995). Finally, HIV-1 gag-specific CTL activity has been demonstrated in an uninfected perinatally exposed child (Rowland-Jones et al. 1993).

In spite of the information supporting the role of CTLs in conferring immunity to infection there is also evidence to the contrary. CTL activity has not been associated with protection of vaccinated primates against viral challenge (Sheppard et al. 1994). An experiment in which expanded autologous CTLs were infused to supplement CTL activity and perhaps lower viral load resulted in an abrupt decline in CD_4^+ cells and rise in viral load. This effect has been tentatively attributed to the selection of *nef*+ viral variants by the expanded population of CTLs. The overall picture is clouded further by the finding (Hulskotte et al. 1995) that macaques exhibiting neutralizing antibody and CTL responses were not protected from experimental challenge with $SIV_{mac32H(J5)}$. Finally, the independent reports (Baier et al. 1995; Cocchi et al. 1995) regarding the CD_8^+ associated chemokines may prove to be a relevant correlate of protection but further studies will be required to define their role in protective immunity. Currently, we are left with the task of assessing vaccine performance with a less than ideal set of measuring devices while the hunt for additional correlates of protection continues. The situation is similar to the lack of an ideal animal model, in that no single measure appears to consistently predict protection. Consequently, the most prudent approach would be to design vaccines which induce the broadest humoral and cellular immunity in an attempt to provide all of the potential correlates of protection.

3.4 Importance of Cytokines and Costimulatory Molecules

Different cytokines have been found to induce and/or suppress replication of HIV-1 depending upon the particular in vitro testing system. Table 6 lists a number of these cytokines and their effects on HIV replication in vitro.

In light of these in vitro observations, a number of investigators have looked at in vivo cytokine levels in various body fluids. A number of the cytokines listed have been found at elevated levels in plasma, lymphoid tissue and even in cerebrospinal fluid.

Cytokines have also been postulated to play a direct role in the ultimate pathogenesis of a variety of infectious diseases. Distinct patterns of cytokine production have been associated with either protective (Th1) or nonprotective (Th2) immune responses (Fig. 2). These patterns have primarily been culled from studies using murine CD_4^+ T cell clones. One theory of HIV pathogenesis was based on a shift from a predominantly Th1 to a predominantly Th2 pattern of cytokine production (Clerici et al. 1993). This theory has not been supported by in vivo data gathered either prospectively or from cross-sectional retrospective studies

Table 6. Observed effects of various cytokines on HIV-1 replication in vivo

Cytokine	Induction	Suppression	Comments
IL-1	+	–	Induction only in macrophages
IL-2	+	–	Induction only in T cells
Il-3	+	–	Induction only in macrophages
Il-4	+	+	Active only in macrophages
Il-6	+	–	Induction only in macrophages
IL-10	–	+	Suppression in T cells and macrophages
IL-12	+	–	Induction only in T cells
IFN-α	–	+	Suppression in T cells and macrophages
IFN-β	–	+	Suppression in T cells and macrophages
IFN-β	+	+	Active in T cells and macrophages
GMCSF	+	–	Induction only in macrophages
MCSF	+	–	Induction only in macrophages
TNF-α	+	–	Induction in T cells and macrophages
TNF-β	+	–	Induction in T cells and macrophages
TGF-β	+	+	Active in T cells and macrophages

IL, interleukin; IFN, interferon; GMCSF, granulocyte/macrophage colony-stimulating factor; MCSF, macrophage colony-stimulating factor; TNF, tumor necrosis factor.

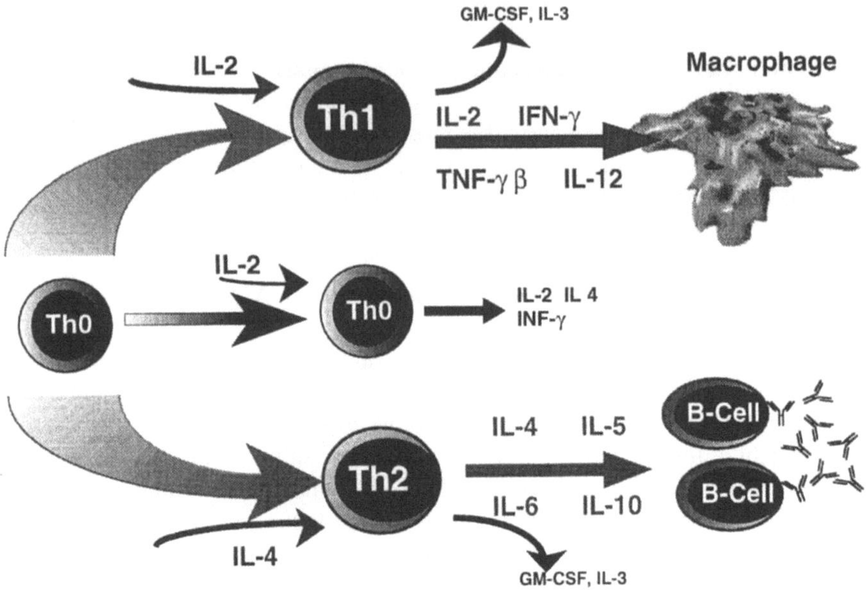

Fig. 2. Helper T cell differentiation into Th1 and Th2 phenotypes

(Graziosi et al. 1994). There is some data which suggest that a predominantly Th0 profile may emerge during the progression of HIV disease (Autran et al. 1995).

In addition to the cytokines listed here, a number of immune cell surface markers are known to play a critical role in the production of effective immune responses. For example, CD_{28} is a marker of naive T cells which acts as a receptor for both B7.1 (CD_{80}) and B7.2 (CD_{86}) molecules on antigen presenting cells (APCs). The interaction between CD_{28} and either B7 costimulatory molecule leads to enhancement of the T cell response. Most importantly, in the context of HIV-1 vaccine design is the finding that T cell stimulation through CD_{28} results in helper T cells which are able to resist infections with HIV-1 (Levine et al. 1996). This observation may ultimately be used to advantage by designers of vaccines for HIV-1.

4 Evidence to Support the Existence of Effective Host Immunity

One of the most significant stumbling blocks in designing an effective HIV vaccine has been the absence of evidence of a protective natural immune response to infection on which to base vaccine design. However, several clinical examples of potentially successful natural responses to HIV-1 infection have now been reported. The most promising evidence to date comes from these rare yet apparently reliable examples of at least partial and in some instances complete control of HIV-1 replication. In each instance native immune responses appear to have successfully thwarted HIV-1 in one way or another. These consist of several groups, listed in subjective order meant to reflect the potency of their presumed immune responses, from least to most impressive:

1. Individuals who remain clinically and immunologically well despite evidence of infection for roughly 10–15 years (Cao et al. 1995; Kirchoff et al. 1995; Pantaleo et al. 1995) including one cohort from a common exposure source (Learmont et al. 1992)
2. A subset of occupationally exposed health care workers with evidence of transient but specific and apparently effective immune response against HIV-1 (Pinto et al. 1995)
3. Two separate reports of children with evidence of neonatal infection who now appear to be virus-free after 3–5 years (Bakshi et al. 1995; Bryson et al. 1995)
4. Individuals with known exposures, (i.e., sexual partners, perinatally exposed children) with no evidence of infection but evidence of an immune response (Rowland-Jones et al. 1993, 1995)

As discussed above, preliminary studies of these individuals have found a number of potentially protective responses, yet not surprisingly, no single explanation can account for the successes even within a single group. However, the fact that these instances have now been identified provides direction for finding better correlates of protection and desirable vaccine characteristics. Most importantly, these results support the notion that immunological control of HIV-1 can be achieved.

5 Alternate Immunization Strategies

The capacity of various vaccine strategies to induce cellular and humoral immune responses are outlined in Table 2. Attenuated viral vaccines are designed to elicit immune responses to relevant disease-associated epitopes without causing symptomatic disease. Historically, attenuated virus has provided the most effective protection against viral diseases with the induction of potent cellular and mucosal responses against diverse viruses such as poliomyelitis, rubella, rubeola and mumps. Of the other approaches, whole inactivated SIV vaccine was shown to induce neutralizing antibodies and initially appeared to protect against heterologous challenge (Johnson et al. 1992). However, protection was later found to be related to alloantigen reactivity between the primate host and the human origin of the inactivated virus (Johnson et al. 1992). In addition, whole inactivated HIV-1 vaccines have not been able to protect from HIV-1 challenge in the chimpanzee model (Schultz et al. 1993). Alternatively, protection from low-dose homologous HIV-1 challenge has been provided by more than one recombinant protein vaccine in the chimpanzee model (el-Amad et al. 1995; Schultz et al. 1993). Protection from intrasubtype B heterologous HIV-1_{SF2} challenge in two chimpanzees vaccinated with gp160 and V3 loop peptides from HIV-1_{MN} (GIRARD et al. 1995) has been reported. This result appeared to correlate with levels of antibodies including those demonstrating viral neutralization. Although this result is encouraging, the inability of recombinant protein/peptide vaccines to induce substantial CTL responses may limit their use in combination or so-called prime-boost strategies. Along these lines, work is ongoing to evaluate the use of certain live recombinant vectors such as canarypox in combination with subunit vaccines (PIALOUX et al. 1995).

Attenuated viral vaccines have been extremely effective in the SIV model although rare yet measurable instances of disease do occur when the vaccine is given to individuals either transiently or permanently immunosuppressed. This drawback has limited current investigations to animal models. The Desrosier's group (DANIEL et al. 1992) has demonstrated that macaques vaccinated (infected) with SIV containing deletions in the *nef* gene resist subsequent challenge with fully virulent SIV. Protection has also been demonstrated against cell-associated infection with SIV (ALMOND et al. 1995). Instances in which HIV-2 infected macaques have resisted infection with SIV have also been reported (PUTKONEN et al. 1991). Chimpanzees may also exhibit protective responses after infection with less virulent strains of HIV-1 (SHIBATA et al. 1995). HIV-1_{DH12} is a strain which appears to be somewhat more virulent in chimpanzees with higher viral burdens and lymphadenopathy observed. Animals already infected with HIV-1_{IIIB} have been shown to be resistant to infection with large inocula of HIV-1_{DH12} strain. However, SIV which behaves as an attenuated agent in adult macaques has shown lethality when given to neonatal macaques via the mucosal route (BABA et al. 1995). In addition, there have been documented cases of HIV-2 infected humans becoming infected with HIV-1 (GEORGE et al. 1992; PEETERS et al. 1992) as well as HIV-1 infected chimpanzees becoming infected with a second unrelated HIV-1 strain (FULTZ et al. 1987). Given the potential risks of live attenuated vaccines for HIV and the evidence against broad protection, additional studies will be required to address both of these important issues.

6 DNA Vaccination as an Approach to Human Immunodeficiency Virus Immunotherapy/Prophylaxis

Although the injection of DNA into tissues was originally reported in the 1950s, the technology has gained more attention in recent years as an apparently inherently safe means of mimicking in vivo protein production normally associated with natural infection. Genetic immunization is dependent upon injection of a nucleic acid sequence directly into a host target tissue (DAVIS et al. 1993; LU et al. 1995; TANG et al. 1992; ULMER et al. 1993; WANG et al. 1993b). The synthesis of specific foreign proteins occurs in the host as in natural infection with either attenuated or wild-type virus. These host-synthesized viral proteins then become the subject of

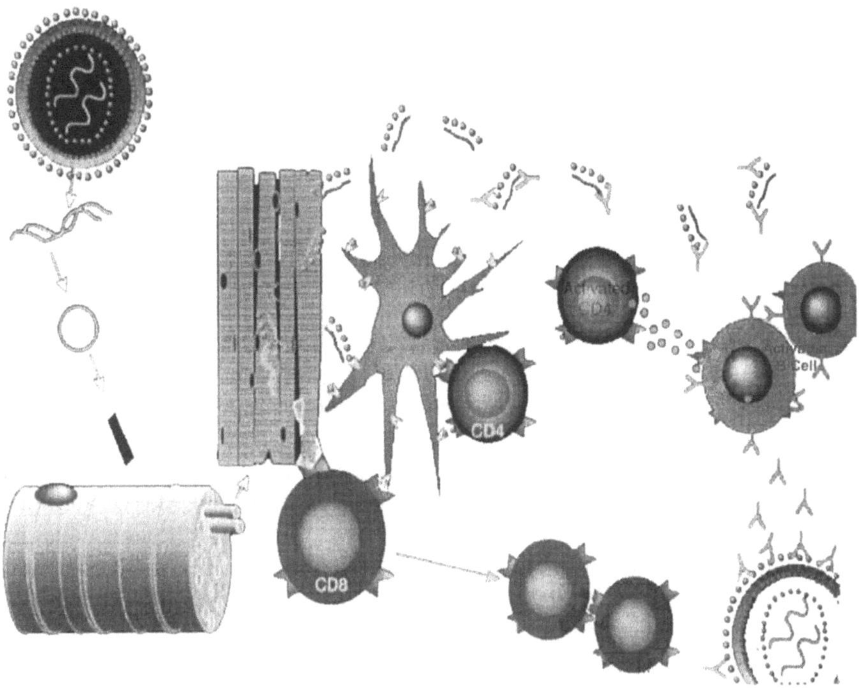

Fig. 3. Proposed mechanisms of immune activation through DNA vaccination

immune surveillance via both the major histocompatibility complex (MHC) class I and class II pathways (WANG et al. 1993b). The immunity generated has included both humoral and cellular responses. These phenomena are summarized in Fig. 3.

Immunization using nucleic acids may afford some potential advantages over better studied strategies such as whole killed or live attenuated virus and recombinant peptide or protein-based vaccines. This newer approach allows specific genes to be expressed in nonreplicating vectors. Designers of these vaccines are able to manipulate the nucleic acid sequences to present all or part of the genome of the organism of interest. Genes which lead to undesired immunologic inhibition or cross-reactivity (autoimmunity) may be either altered or deleted altogether (AYYAVOO et al. 1997).

Once these vaccine expression cassettes are taken up into cells their products may be presented as specific epitopes to the host immune system. In this way, genes which encode important immunologic epitopes can be included while those that confer pathogenicity or virulence can be excluded. This flexibility can be exploited to include sequences of multiple epitopes within a protein or across divergent sequences from different strains. This may be achieved within a single construct as opposed to equally safe inactivated vaccines which require multivalent structures to achieve the same goal. Vaccines using inactivated virus also do not elicit strong cytotoxic T cell-mediated cytolysis (CTLs). In DNA vaccination the host would provide posttranslational modifications to present antigens which faithfully reproduce native conformations, thus avoiding a potential problem of recombinant protein-derived vaccines which may not achieve native tertiary structure.

These genetic constructs should be able to elicit CTLs by the same mechanism as live attenuated vaccines while avoiding the risk of transmission of revertant virus. Helper T cell responses in association with class II MHC antigens will also be provided as the vaccine genes are transcribed and translated and then presented for immune surveillance. Peptides and/or proteins created from vaccine and genes could provide specific targets for antibody-mediated responses which induce neutralization, inhibit syncytium formation and provide protection at mucosal surfaces. All of these responses have been demonstrated using nucleic acid-based vaccines and efforts are ongoing to further these ideas through a variety of approaches.

Although the need for alternative immunization approaches may be keenest for certain viral diseases, investigators are currently studying this strategy for preventing infection with bacteria, mycobacteria, parasitic helminths and for therapy of cancer. The viral diseases studied include bovine hepatitis, hepatitis B, human T lymphotropic (HTLV-1), hepatitis C, rabies, influenza as well as HIV-1 and the related retroviruses. Our group has initiated studies of DNA vaccination as a methodology for producing broad immunity against HIV-1 (Boyer et al. 1996; Wang et al. 1993a, b, 1994b, 1995a, b c). Given the unique challenges found in HIV-1 vaccine development several additional investigators have begun to study this new technology as it pertains to this difficult viral target. These efforts have produced several encouraging results which would appear to warrant further investigation.

6.1 In Vitro Data

Wang demonstrated expression of HIV-1 envelope proteins in TE671 cells transfected by a plasmid engineered with an insert containing *tat*, *rev* and gp160 sequences from an HxB2 isolate of HIV-1 (pM160) (Wang et al. 1993b). Cells found to express gp160 protein were isolated by their specific binding to anti-gp120 monoclonal antibodies and their subsequent protein expression was determined by western blot analysis. Follow-up studies (Wang et al. 1993a, 1995c) reproduced similar results in response to constructs encoding *tat*, *rev*, *vpu* and gp160 sequences from other North American isolates of HIV-1 (Z6, MN). The expression of a plasmid which contains the *gag*, *pol* and *rev* sequences of HIV-1_{HxB2} by measurement of p24 production and by ELISA and immunofluorescence/flow cytometry assays has also been reported (Coney et al. 1994).

6.2 In Vivo Data

6.2.1 Small Mammal Studies of DNA Vaccination

In experiments related to those described in Sect. 6.1, plasmids were injected intramuscularly 24 h after the same site was facilitated with bupivacaine hydrochloride. Bupivacaine treatment is thought to stimulate myocyte division and can induce muscle regeneration at high concentrations (THOMASON et al. 1990), thus increasing nucleic acid uptake into the myocyte (DANKO et al. 1994) and thereby increasing expression and ultimately immunity (WANG et al. 1993b). In an initial study, groups of ten mice received a series of four injections each separated by 2 weeks. These genetic vaccines were found to generate multifaceted immune responses in mice after facilitated injection with the amide anesthetic. Sera obtained from eight of the ten pM160-injected mice reacted with recombinant gp160 protein as determined by ELISA. In addition, neutralizing antibodies against both homologous and heterologous isolates were induced. The immune response mapped to several diverse regions of gp160 including: (a) $CD_4{}^+$ binding domain of HIV-1_{HxB2}; (b) V3 loop peptide of HIV-1_{BRU}; (c) V3 loop peptide of HIV-1_{MN}; (d) V3 loop peptide of HIV-1_{Z6}; (e) the fusogenic region of gp41, (F560 peptide); and (f) the immunodominant region of HIV-1_{BRU} (Fig. 4). Seroconversion was also demonstrated in 100% of mice after a single immunization with the gag/pol construct described above (CONEY et al. 1994) at dilutions of at least 1:1000. Antibodies to Rev protein were demonstrated at dilutions of 1:10000.

In a related study (WANG et al. 1994b), mice were subjected to intramuscular injection with a construct encoding gp160 of HIV-1_{Z6} or vector alone followed by excision of the injected muscle 72 h later. Muscle sections were treated with a pool of murine anti-HIV-1 gp160 and gp41 monoclonal antibodies and specific reactivity was determined by indirect immunohistochemistry with an HRP-conjugated detection system. Reactivity to both gp160 and gp41 was observed along needle tracks through the muscle, thus demonstrating that gp160-specific antigen can be expressed in myocytes inoculated with gene constructs.

The antisera of all pM160 immunized animals neutralized HIV-1_{IIIB} in vitro compared to pre-immune sera. Post-immune sera (but not pre-immune) demonstrated in vitro neutralization activity at dilutions up to 1:1280 for HIV-1_{IIIB} and HIV-1_{Z6} viral stocks. The pM160-MN construct demonstrated neutralization of homologous isolates at titers up to 1:320 but also showed lower yet measurable neutralization to the heterogeneous HIV-1_{Z6} isolate (WANG et al. 1995c). In a related assay, antiserum from mice inoculated with gp160 constructs was also found to inhibit syncytium formation at a dilution of 1:200. Another indication of the humoral response elicited by vaccination was obtained by measuring the ability of antisera to inhibit gp120 binding to CD_4^+ bearing T cells. Antisera diluted at 1:5 and 1:15 were shown to inhibit gp120 binding to CD_4^+ SupT1 cells by 22% ± 2% using flow cytometry. These findings support the contention that DNA inoculation is capable of inducing conformationally relevant antigens which may be able to elicit cross-reactive responses. Neutralization of homologous isolates has also been

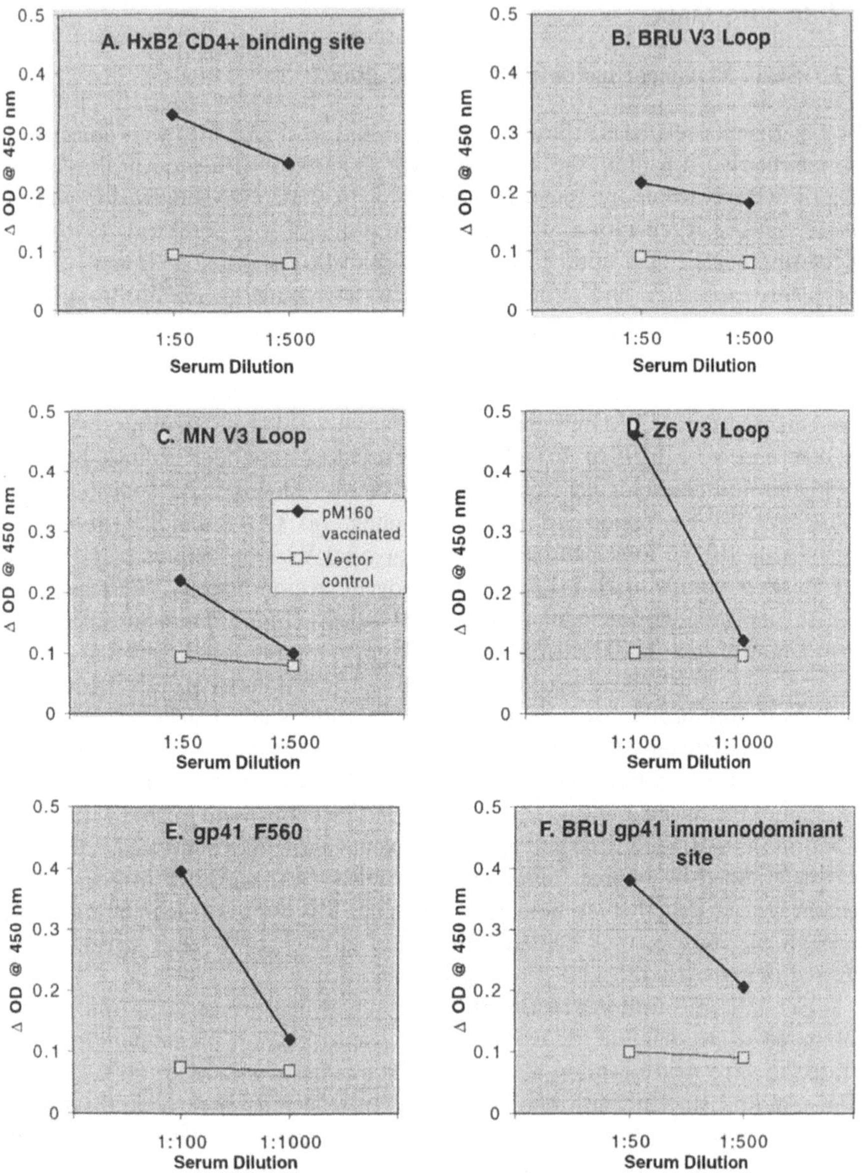

Fig. 4A–F. Antibody binding to peptides representing gp160 epitopes for pM160 vaccinated animals vs unimmunized controls

reported after immunization with constructs based on the HIV-$1_{NL4\text{-}3}$ isolate in the presence of relatively low anti-ENV IgG titers (Lu et al. 1995).

Finally, immunoglobulin isotyping studies of the gp160-specific antibodies elicited by envelope DNA vaccination revealed a predominance of IgG isotypes with 19% IgG1, 51% IgG2, 16% IgG3, 10% IgM and 5% IgA responses measured. These results are consistent with a secondary immune response suggesting helper T cell stimulation and anti-viral activity through nucleic acid-based immunization. Furthermore, this immunoglobulin isotype distribution is a hallmark of a Th1 type response (WANG et al. 1993b).

The Z6 envelope construct induced seroconversion against gp160 of increasing numbers of mice with each boost, with 95% responding by the fourth injection. Okuda and colleagues (OKUDA et al. 1995) also reported a similar boosting effect in a rabbit injected with their plasmid combination. They also found that the second *rev* construct could augment anti-ENV responses above what was found when the *env* construct was injected alone. This may represent intracellular cooperation between exogenously delivered gene products. Since pM160-Z6 also contained sequences coding for Tat and Rev proteins additional assays were performed which demonstrated specific antibody formation against these proteins by ELISA. Antibody responses were also demonstrated to more than one antigen (transient anti-ENV and persistent anti-p24) using a construct which produces noninfectious viral particles (LU et al. 1995). These results independently illustrate the ability of DNA-based vaccines to elicit specific immune responses to multiple viral antigens using a single construct. The incorporation of multiple genes in a single DNA-based vaccine could potentially further expand the immunologic repertoire seen with single gene constructs.

The longevity of anti-gp120 immune responses has been ascertained using the HIV-1_{MN} envelope construct. Figure 5 depicts the optical density observed at

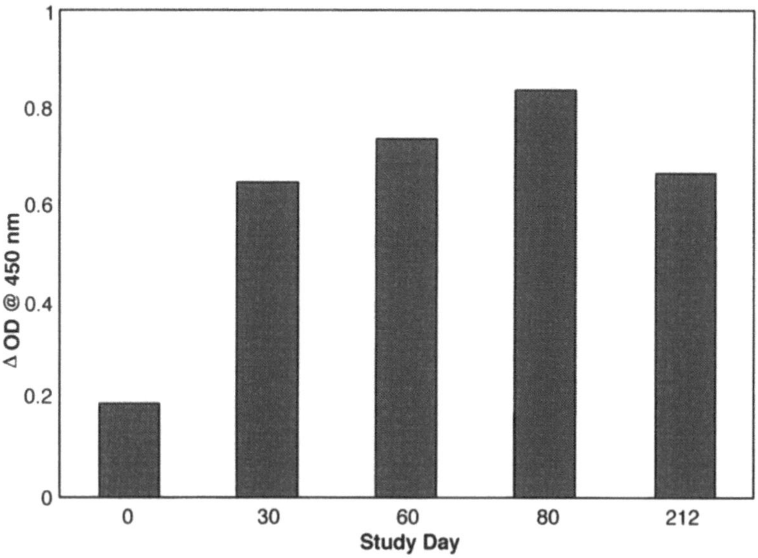

Fig. 5. Longevity of antibody response to HIV-1(MN) gp120 in pM160-MN immunized mice (1:100 serum dilution)

various days post-inoculation with a single dose of vaccine. This result illustrates that specific responses may be maintained up to one third of the murine lifespan through immunization with DNA constructs.

Investigators (OKUDA et al. 1995; WANG et al. 1994b) have demonstrated the presence of specific RNA in the muscle of mice simultaneously injected with DNA from several different plasmids encoding sequences of HIV-1. Immunohistochemical staining performed on muscle sections injected with the plasmids showed reactivity when stained with anti-HIV-1 antibody. In addition, antisera from rabbits immunized with this plasmid combination were also analyzed by an HIV growth inhibition assay utilizing p24 production as a growth indicator. Antisera from rabbits immunized with the same constructs were also found to inhibit p24 production by HIV-1-infected human PBMCs when compared to pre-immune sera.

Splenocytes from envelope inoculated mice were compared to vector inoculated controls for their ability to proliferate in response to specific stimulation with recombinant gp120 (WANG et al. 1993b). Subsequent studies (CONEY et al. 1994) have broadened these results with other constructs in response to gp120, gp160, gp41 external domain and several peptides of gp120 and gp41. Representative murine proliferative as well as CTL responses to HIV-1 gag/pol DNA vaccination are depicted in Fig. 6 (Bagarazzi, unpublished data).

The induction of CTLs after vaccination with envelope DNA constructs has also been reported (WANG et al. 1993b). Percent-specific lysis was measured to be 32

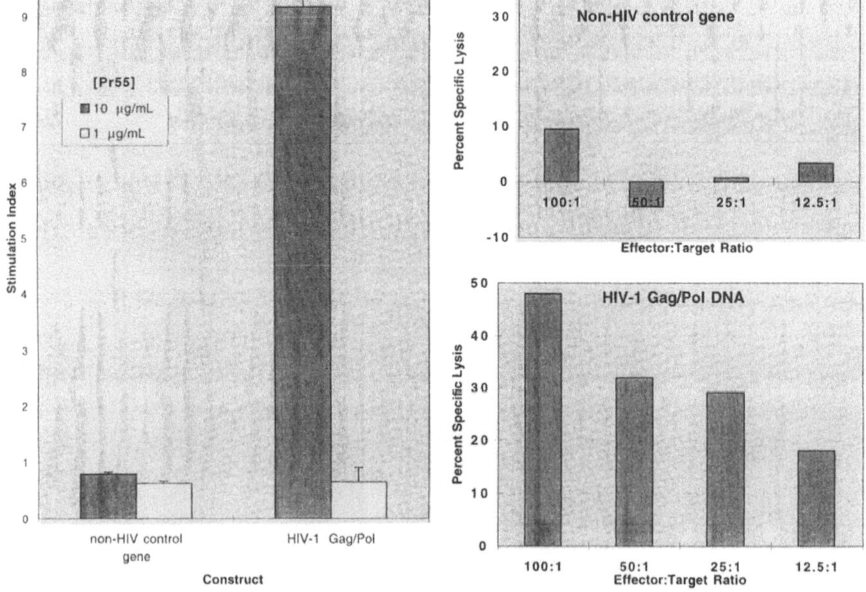

Fig. 6A, B. Cellular responses of BALB/c mouse splenocytes to DNA vaccination with HIV-1 gag/pol constructs. **A** Proliferation in response to HIV-1 Pr55; **B** CTLs in response to gag/pol targets

and 13 at effector:target (targets infected with VPE16 vaccinia virus encoding HIV-1 gp160) ratios of 30:1 and 10:1, respectively, compared to background levels in response to immunization with recombinant envelope protein. Constructs designed from the HIV-$1_{NL4\text{-}3}$ isolate were also shown to elicit CTL activity (Lu et al. 1995). The successful production of CTL responses directed against a V3 loop peptide of HIV-1 gp120 has also been demonstrated (Haynes et al. 1994). Specific CTLs were reported after a single epidermal immunization with a HIV-1 gp120 construct delivered via the Accell gene delivery system. The level of interferon (IFN)-γ and interleukin (IL)-4 released from stimulated splenocytes revealed IFN-γ levels paralleled CTL activity while IL-4 levels increased with the number of immunizations. Consequently, HIV-1 constructs have been successfully expressed by different means to different tissues, each method exhibiting both humoral and cellular immune responses. However, the cytokine profiles observed suggest that immunization route may influence the type of immune response elicited by DNA vaccination.

Measurements of footpad swelling in response to specific antigen injection were evaluated as a means of determining delayed type hypersensitivity (DTH) reactions in mice immunized with other HIV-1 constructs (Okuda et al. 1995). Although such responses are far from standardized and difficult to interpret, it appeared that satisfactory DTH responses were measured after a single dose of the plasmid combination. Induction of antigen-specific lymphocyte proliferation was also assessed in mice injected with plasmid constructs. Proliferation was seen in response to both homologous peptides (III_B) and to a lesser extent with a peptide from a heterologous E (Thai A) strain. CTL activity in response to V3 PND peptide was also assessed in these animals with vaccine recipients demonstrating approximately 25% specific lysis at E:T ratio of 100:1 with correspondingly negligible activity among controls. The combination of the above independent findings supports the assertion that nucleic acid-based vaccines are indeed capable of inducing relevant cellular responses to HIV-1 antigens in this small mammal model.

Murine studies of potential HIV-1 vaccine strategies are limited to demonstrating immunogenicity since the animals are not susceptible to infection with HIV-1 nor the subsequent immunodeficiency. A lymphoma rejection model was developed to assess the ability of nucleic acid-based immunization to protect against relevant cellular challenge (Wang et al. 1994a). A clonal population of lymphoma cells (SP2/0) was transfected with HIV-1 gp160 envelope and shown to express the transfected antigen in association with H-2^d class I MHC antigens. All naive and vector immunized control mice subsequently challenged with gp160 expressing SP2/0 cells developed tumors within 13 days and succumbed within 9 weeks. In contrast, all eight animals previously immunized with the HIV-1_{Z6} envelope construct survived the same lethal challenge (Fig. 7) and six never developed tumors over a 6 month follow-up period. Since cellular immunity is a critical component of tumor rejection one can project that successful protection is likely a direct result of cell-mediated responses directed against the gp160 antigen expressed in the SP2/0 cells.

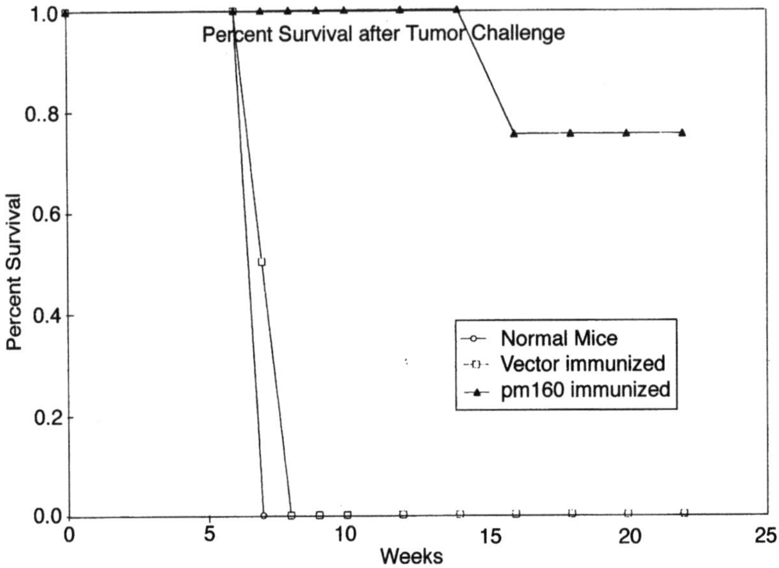

Fig. 7. Survival of mice challenged with syngeneic SP2/0-gp160 tumor cells after vaccination with nothing (*open circles*), vector (*open squares*) or pM160 (*closed triangles*) DNA vaccine

As mentioned in Sect. 3.4, murine and primate experiments have been performed in an effort to engineer specific immune responses through the co-administration of a number of cytokine and/or costimulatory molecule genes with specific viral genes (Kim et al. 1997a, b). The data suggests that the activity of genetic immunization may be affected in a number of ways. Specifically, IL-12 vectors increase proliferative and specific anti-HIV-1 CTL activity while diminishing antibody generation in mice (Kim et al. 1997a, b, c). Additionally, co-administration of granulocyte/macrophage colony-stimulating factor (GMCSF) sequence vectors with HIV-1 DNA vaccines increased T cell proliferative and antibody responses (Kim et al. 1997a, b, c). In a related study, the costimulatory molecules CD_{80} and CD_{86} were co-injected with specific HIV-1 DNA vaccines. Both T cell proliferative

Table 7. Observed in vivo effects of cytokine and costimulatory molecule expression cassette co-administration on immune responses induced by HIV-1 expression cassettes

Cytokine	Antibody production	T cell proliferative responses	CTL responses
Interleukin-12	decreased	increased	increased
GMCSF	increased	increased	unaffected
CD_{80} costimulatory molecule	unaffected	slightly increased	unaffected
CD_{88} costimulatory molecule	unaffected	increased	increased

CTL, cytotoxic T lymphocyte; IL, interleukin; GMCSF, granulocyte/macrophage colony-stimulating factor.

and CTL responses were augmented with CD_{86} in a manner similar to IL-12 coadministration, while CD_{80} administration induced a modest increase in proliferative responses but had little effect on CTLs. However, CD_{86} coadministration did not affect humoral responses. These results are consistent with known cytokine function and may represent enhanced immunogenicity by increasing cytokine activity at the sites of immunization (Table 7).

6.2.2 Small Primate Studies of DNA Vaccination

The pM160-Z6 construct described in Sect. 6.1 (WANG et al. 1995b) was also used to immunize four cynomolgus monkeys. Sera from three of the four monkeys reacted with HIV-1_{HxB2} rgp160 and rgp120 and an external immunodominant domain of gp41 (amino acids 562–641) at dilutions of 1:400 or greater. These three animals also produced antibodies in response to HIV-1_{SF2} gp120 and a nonglycosylated gp120 peptide at dilutions of at least 1:400. The fourth animal was tested for reactivity to HIV-1_{HxB2} rgp120 and a titer of 1:400 was obtained. Epitope mapping was performed on serum samples obtained after the fourth immunization against a series of twelve different peptides from the HIV-1_{MN} envelope sequence. Two of the three immunized animals consistently reacted with each peptide, while one only reacted to the first two peptides at the NH_2-terminal. In addition, sera obtained from the same two of three monkeys demonstrated in vitro viral neutralization activity against homologous HIV-1_{Z6} and heterologous HIV-1_{MN} isolates at dilutions of up to 1:640 and 1:320, respectively (WANG et al. 1995b). A similar correlation between peptide mapping and neutralization has previously been reported in humans (UGEN et al. 1992), as this may be an important feature of this vaccination strategy.

Other constructs (OKUDA et al. 1995) based on the envelope gene of HIV-1_{IIIB} were also found to elicit specific anti-HIV-1 humoral responses when administered to macaques. Neutralization and syncytium inhibition assays performed on these animals showed significant titers in both assays against different HIV-1 strains. In keeping with reports from investigators using other immunization strategies, the neutralization of homologous HIV-1_{IIIB} virus was obtained at higher dilutions than when infection was carried out with heterologous HIV-1_{SF2} or HIV-1_{Gun1} (strain found in Japan).

Cellular responses were also confirmed in the cynomolgus macaques immunized with the envelope construct. All four pM160-Z6 immunized macaques showed excellent proliferation in response to stimulation with gp160 and gp120 from HIV-1_{HxB2} (WANG et al. 1995b). This would suggest that enough epitopes were presented in association with class II MHC antigens to induce a significantly broad response in this outbred animal model. Since the construct also contains *rev* and *tat* sequences, PBMCs from one of the animals were assayed for proliferation in response to Rev and Tat protein stimulation. There was no response to Tat but responses to Rev equaled or surpassed those seen against the envelope proteins. There appears to be little antigenic interference in this model.

The same pM160-Z6 construct was also shown to elicit specific CTL activity in response to stimulation with gp160-expressing vaccinia in a macaque at E:T ratios of 50:1 and greater (CONEY et al. 1994). CTL responses were also noted in a second animal in response to gp120 as well as gp160 but there was no activity seen in response to a third truncated peptide that expresses only the first 200 amino acids of gp120. This suggests that recognized epitopes were distributed toward the COOH-terminal of gp120 and/or gp41. Two animals showed significant CTL activity against HIV-1_{IIIB} envelope targets, particularly following the fourth immunization with a 100 μg dose of plasmid (WANG et al. 1995b). CTL lysis was not seen from one of the immunized animals and the unimmunized control. This constituted the first demonstration of effector CTL induction in a primate through a nucleic acid-based vaccine.

CTL activity in response to V3 PND peptide was assessed in macaques immunized with the pCMVIIIB/REV vaccine (OKUDA et al. 1995). Vaccine recipients demonstrated CTL activity with percent-specific lysis varying from 35 to 15 at E:T ratios of 100:1 to 12.5:1, respectively. CTL activity among controls was negligible. The combination of the above independent findings would support the assertion that nucleic acid-based vaccines are indeed capable of inducing relevant cellular responses to HIV-1 antigens.

Since broad humoral and cellular responses had been measured in response to genetic vaccination in these small primates the next step was to attempt to demonstrate protection from viral challenge. The first demonstration of primate protection from viral challenge using this technology was accomplished using a SIV/HIV-1 chimera (S/HIV-1) (BOYER et al. 1996). One of four cynomolgus macaques vaccinated with envelope plasmid DNA was protected from intravenous challenge with 50 $TCID_{50}$ of a S/HIV-1_{HxB2} chimeric (i.e., heterologous envelope) virus stock. Viral load was somewhat limited in two additional animals which did not seroconvert to SIV core proteins. The fourth animal was clearly infected as demonstrated by both measurable viral load and seroconversion. This study demonstrated that DNA vaccination could lead to control of retroviral replication in vivo. Clearly however, additional studies showing protection in a higher percentage of animals are an important goal of further challenge studies.

DNA vaccines encoding genes of SIV and the closely related HIV-2 have also been studied in the rhesus macaque model. In one study, CTL activity of macaques vaccinated with a combination of DNA vaccines encoding all or part of both envelope (four different plasmids) and gag (one plasmid) genes of SIV were evaluated (YASUTOMI et al. 1996). DNA was administered on six separate occasions both intravenously and intramuscularly as well as by a gene gun using gold beads with cumulative DNA doses of 16–21 mg. CTLs were assessed 1, 4 and 7 weeks after each of three clusters of two vaccinations (separated by 2 weeks). In one group of four animals vaccinated by all three routes, envelope-specific CTLs (greater than 10%) were measured in each animal by 1 week after the second cluster. Two of these four animals also exhibited CTLs specific to gag. The three animals vaccinated using only the gene gun exhibited only envelope specific CTLs. The CTLs measured were shown to be populations of CD_8^+ lymphocytes with

MHC class I restriction. These monkeys were challenged with SIV_{mac251} after their sixth immunization but protection was not seen, possibly some attenuation in viral load was seen in comparison to animals vaccinated with vector alone (Lu et al. 1996). Neutralizing antibodies were measured in the monkeys although only transiently in comparison to CTLs which appeared to be more durable.

Our group has also studied the immunogenicity of DNA vaccines encoding HIV-2 when administered to mice of different haplotypes (Agadjanyan et al. 1997). Antibodies to both HIV-2 envelope as well as the related SIV envelope were measured. These included the demonstration of neutralizing capacity against the ROD isolate of HIV-2. The murine splenocytes also exhibited specific proliferative responses against both SIV and HIV-2 proteins. CTLs specific to the HIV-2 envelope were also measured, although only after coadministration of an IL-12 expression vector in the previously described manner (Kim et al. 1997a, b, c). These results show not only that DNA vaccines are potentially effective against yet another viral target but that these responses are also cross-reactive against related yet distinct viruses. In addition to the above studies in small mammals and primates, a number of recent studies have now evaluated the use of nucleic acid vaccines in the chimpanzee model.

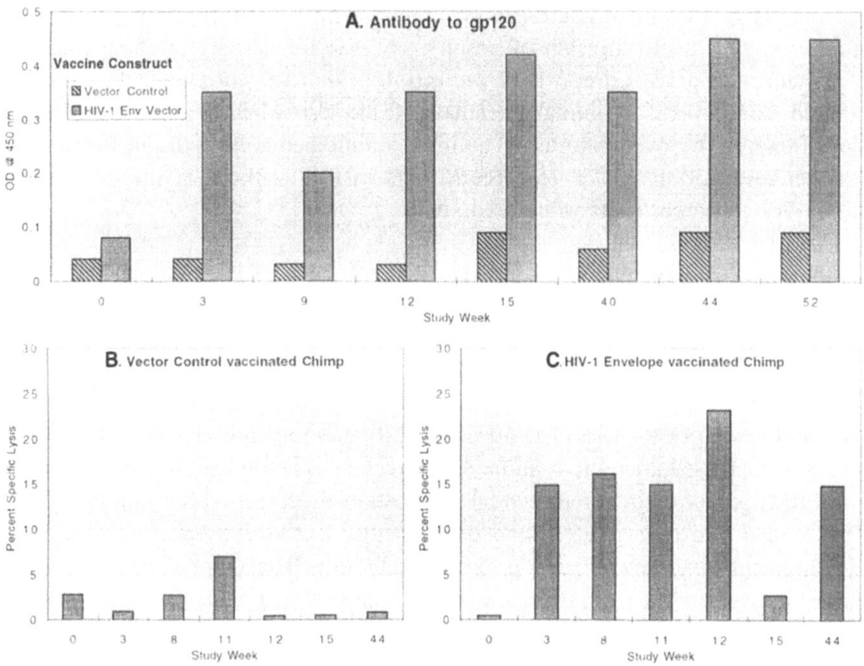

Fig. 8A–C. Chimpanzee immune responses after vaccination with HIV-1 envelope DNA vaccine. **A** antibodies to gp120 by ELISA. CTLs to envelope targets after vaccination with vector control DNA (**B**) or HIV-1 envelope DNA (**C**)

6.2.3 Chimpanzee Studies of DNA Vaccination

Three chimpanzees inoculated with an HIV-1_{MN} envelope construct developed both cellular and humoral immune responses to specific envelope antigens (Boyer et al. 1995). Antibody responses to cyclic V3 loop peptide of HIV-1_{MN} were measured for each animal by the third boost, in comparison to a vector immunized animal which never demonstrated measurable antibody (Boyer et al. 1996). Vaccination also induced each animal to develop antibodies to gp120 as can be seen in Fig. 8a. (Boyer et al., in press). These animals were later immunized with a gag/pol construct and one developed antibody responses to Pr55 core protein. Serum from all three immunized animals was able to neutralize HIV-1_{MN} in vitro.

CTL activity was measured in two of the three immunized animals. One animal developed activity to both Gag/Pol and Env targets (see Fig. 8b) and the other showed activity against Gag/Pol alone. Two animals and the control animal were subsequently challenged intravenously with a high dose (250 chimpanzee ID_{50}) of a heterologous stock of HIV-1_{SF2}.

Both chimpanzees challenged in this manner were found to be protected (Boyer et al., 1997). The control vector-immunized animal became infected by the identical challenge. This result represents the first example of protection from viral challenge by a genetic vaccine in the chimpanzee model. This approach using both core and gp160 constructs was able to prevent infection from a strain heterologous to the envelope construct. Interestingly, of the two animals which were protected, one displayed substantial neutralization activity while the other had superior CTL responses. This demonstration of prophylactic vaccine activity in chimpanzees is consistent with previous reports of protection in that the correlates of protection differed from animal to animal. It remains to be seen whether antigenic diversity can ultimately be overcome using multiple simultaneous antigens in humans or whether several different envelope constructs with or without additional core or regulatory gene constructs will be required.

6.2.4 Chimpanzee Studies of Immunotherapy with DNA Plasmid Immunogens

DNA vaccination has been evaluated for its utility as a putative immunotherapeutic agent. An adult chimpanzee already infected with HIV-1 was vaccinated with the same envelope construct given to the naive chimpanzees mentioned above (Boyer et al., in press; Ugen et al. 1997b). Antibodies to peptides within both gp120 and gp41 increased after vaccination. Measurements of viral load by both RT-PCR and PBMC coculture methods exhibited a sustained decrease after immunization. This animal had been infected and demonstrated consistently positive viral load measurements over several years prior to vaccination. This animal appears to have partially cleared virus from its intravascular compartment while exhibiting at least a humoral response to HIV-1. The reduction in measurable virus has persisted for at least 12 months after vaccination. The encouraging results of the two aforementioned studies coupled with the lack of any significant demonstrable toxicity in the vaccinated chimpanzees would suggest that DNA vaccines should be explored in humans for the control and treatment of HIV-1 infection. A Phase I trial of the same envelope construct has begun in humans already infected with HIV-1.

6.2.5 Human Studies of DNA Vaccination

The trial mentioned above represented the first use of a nucleic acid-based vaccine in humans and has enrolled 20 HIV-1 seropositive volunteers with more than 500 CD_4^+ lymphocytes/ml. Patients in the trial receive three injections each separated by 10 weeks with escalating dosage (three dosage groups of five subjects) of envelope vaccine. Preliminary results reveal no clinical or laboratory adverse effects measured in all three dosage groups (30, 100, 300 μg) (MacGregor et al. 1996b).

Antibody responses to envelope proteins and peptides of individuals receiving the 100 μg dose have been shown to increase after the onset of immunization (MacGregor et al. 1996a; Ugen et al. 1997a). Assays of CTL activity have revealed increases in percent-specific lysis of envelope constructs in several vaccine recipients although the durability of these responses has not been measured. Viral load measurements to date have not shown any clear pattern of either increase or decrease. Similarly, measurements of lymphocyte subsets by FACS analysis has shown neither positive or negative effects on at least the CD_4^+ or CD_8^+ populations of vaccinated individuals. These preliminary results demonstrate that the injection of even relatively low doses of a single immunogen DNA vaccine is capable of augmenting both humoral and cellular immune responses in humans.

A Phase I trial using the same envelope vaccine has now gotten under way among human volunteers seronegative for HIV-1. Additionally, a second Phase I trial to evaluate a gag/pol DNA vaccine has also commenced. These studies represent the first opportunity to demonstrate de novo immune responses as a result of inoculation of plasmid DNA into human subjects. Data analysis has just begun on these interesting trials. Taken together, these studies have been responsible for pioneering DNA vaccines for clinical use in humans.

7 Perspectives

The field of nucleic acid-based immunization has undergone rapid and formidable change in just the last 5 or 6 years from the publication of four seminal papers. The field has developed to the point where genetic vaccines are currently under development for virtually every type of human infection, from viruses to mycobacteria and plasmodia. Protection from infection has already been demonstrated in more than one disease model and now several human trials are ongoing simultaneously. The development of these vaccines for the prevention and treatment of disease due to HIV-1 has advanced in step with the technology in general. Several firsts in the field of DNA vaccination have occurred using vaccines against HIV-1. Antibody and cellular (proliferation and CTL) responses have now been demonstrated in everything from mice to humans. Disease prevention was shown in a murine tumor challenge model, a macaque model using S/HIV-1 chimeras, and now in a chimpanzee model. The therapeutic use of these vaccines to augment existing immune responses has been demonstrated in chimpanzees and humans.

These intriguing developments have occurred despite the fact that the mechanism(s) of genetic immunization remain to be thoroughly elucidated. Different groups have demonstrated relevant immunity using a variety of doses, facilitators, routes, schedules and immunization strategies. The fine specificity and prevailing character of the immune response is also the subject of debate. The further elucidation of any or all of these questions will surely provide more focused and consequently even more rapid development of the technology. In addition, the approaches to co-immunization have varied from the use of multiple separate plasmids to multiple genes per plasmid. The search for an optimal approach is still very much ongoing.

The overwhelming majority of the work on DNA vaccines for HIV-1 has been accomplished using single gene constructs without the answers to the questions mentioned above. Our group has approached the formidable problem of designing a vaccine against HIV-1 with the intent to incorporate multiple immunogens representing a variety of genes into a single vaccine with the ability to induce broad-based immunity (see Fig. 9). We feel that virtually the entire genome of HIV-1 may ultimately be incorporated into a successful vaccine. Those genes which are naturally nonpathogenic may be used as wild-type sequences to take advantage of any important immunogenic epitopes. Genes which are involved in the pathogenesis of

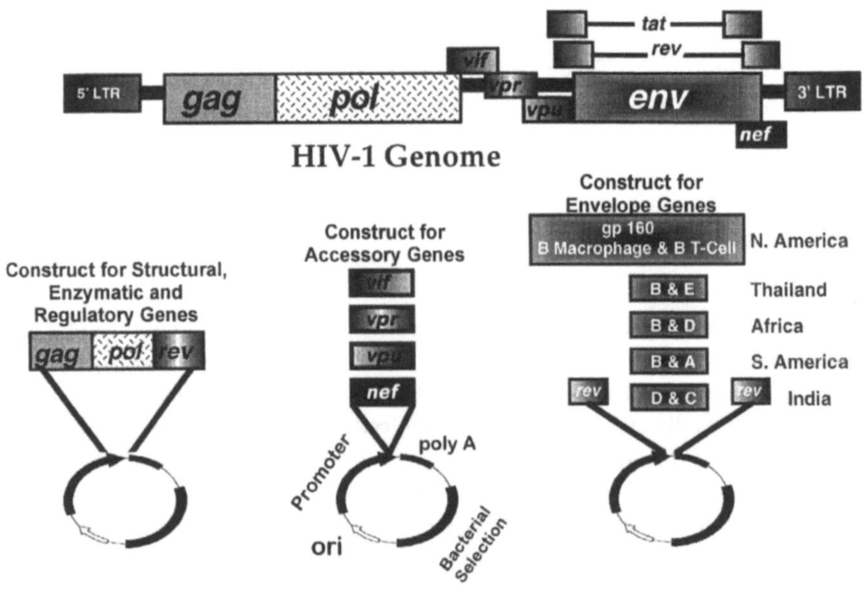

Fig. 9. Design of a prophylactic vaccine to HIV-1

HIV-1 infection and the progression to AIDS must be altered to render them nonfunctional while preserving as many immunogenic epitopes as possible. There is also the exciting prospect for developing multiple envelope subtype constructs in an attempt to address the antigenic diversity of HIV-1 for use internationally. Lastly, coadministration with the cytokine and costimulatory molecule sequences affords the opportunity to augment specific humoral or cellular responses with constructs encoding appropriate epitopes. Admittedly, this is but one strategy which may be used and if a successful vaccine is developed it will likely incorporate the ideas of a wide variety of investigators.

Much has been learned in the pursuit of a vaccine for HIV-1 yet no clear consensus has been reached on whether infection and/or disease will ever be controllable through immunization. Even the most optimistic interpreters of the in vitro and animal data must still question the clinical relevance of any findings. The most pessimistic (Hilleman 1992) make convincing arguments that the challenges facing HIV vaccinologists may prove to be too formidable. Past successes are even used as evidence that HIV-1 will not prove amenable to vaccines because of its many unique properties compared to vaccine-responsive viruses. Optimists take solace in the isolated cases of presumed viral clearance by host immune responses and the chemotherapeutic successes found with combinations of anti-retroviral agents. The recent achievements in the area of DNA vaccines suggest a novel path for scientific investigation to formally test the ability of a combination vaccination approach for HIV-1. It is likely that these studies will significantly change our thinking and approaches to the design of vaccines now as well as in the forseeable future. Given the many approaches investigated and still more suggested, a lack of imagination has not been a serious shortcoming of the field of HIV research. Perhaps the answers lie in some combination of vaccine(s), traditional chemotherapy, and even gene therapy. One can lament that only time will tell, but, all of us in the field must never forget that the wait is literally killing people, so time is of the essence.

References

Agadjanyan M, Trivedi NN, Levine W, Kim JJ, Kudchodkar S, Boyer JD, Bennett M, Ugen KE, Weiner DB (1997) HIV-2 expression vectors generate cross-strain humoral and cellular anti-HIV-2 and anti-SIV immune responses. In: Brown F, Chanock RM, Ginsberg HS, Norrby E (eds) Vaccines 97: molecular approaches to the control of infectious diseases. Cold Spring Harbor, pp 131–136

Agy M, Frumkin LR, Corey L, et al. (1992) Infection of Macaca nemestrina by human immunodeficiency virus type 1. Science 257:103–106

Almond N, Kent K, Cranage M, Rud E, Clarke B, Stott EJ (1995) Protection by attenuated simian immunodeficiency virus in macaques against challenge with virus-infected cells. Lancet 345:1342–1344

Anderson D, Politch JA, Martinez A, Van Voorhis BJ, Padian NS, O'Brien TR (1992) White blood cells and HIV-1 in semen from vasectomized seropositive men. Lancet 338:573–574

Autran B, Legac E, Blanc C, Debre P (1995) A Th0/Th2-like function of CD4+/CD7- T helper cells from normal donors and HIV infected patients. J Immunol 154:1408–1417

Ayyavoo V, Nagashunmugam T, Boyer JD, Sundarasamy M, Fernandes LS, Lee P, Lin J, Nguyen C, Chattergoon MA, Goedert JJ, Friedman H, Weiner DB (1997) Development of genetic vaccines for pathogenic genes: construction of attenuated Vif DNA immunization cassettes. AIDS (in press)

Baba T, Jeong YS, Pennick D, Bronson R, Greene MF, Ruprecht RM (1995) Pathogenicity of live attenuated SIV after mucosal infection of neonatal macaques. Science 267:1820–1825

Baier M, Werner A, Bannert N, Metzner K, Kurth R (1995) HIV suppression by interleukin-16. Nature 378:563

Bakshi SS, Tetalli S, Abrams EJ, Paul MO, Pahwa SG (1995) Repeatedly positive human immunodeficiency virus type 1 DNA polymerase chain reaction in human immunodeficiency virus-exposed infants. Pediatr Infect Dis J 14:658–662

Barrett N, Eder G, Dorner F (1991) Characterization of a vaccinia-derived recombinant HIV-1 gp160 candidate vaccine and its immunogenicity in chimpanzees. Biotech Therapeut 2:91–106

Berman P, Gregory TJ, Riddle L, Nakamura GR, Champe MA, Porter JP, Wurm FM, Hershberg RD, Cobb EK, Eichberg JW (1990) Protection of chimpanzees from infection by HIV-1 after vaccination with recombinant glycoprotein gp120 but not gp160. Nature 345:622–625

Bleul C, Farzan M, Choe H, Parolin C, Clark-Lewis I, Sodroski J, Springer TA (1996) The lymphocyte chemoattractant SDF-1 is a ligand for LESTR/fusin and blocks HIV-1 entry. Nature 382:829–832

Borrow P, Lewicki H, Hahn BH, Shaw GM, Oldstone MB (1994) Virus-specific CD_8^+ cytotoxic T-lymphocyte activity associated with control of viremia in primary human immunodeficiency virus type-1 infection. J Virol 68:6103–6110

Boyer J, Wang B, Ugen KE, Srikantan V, Gilbert L, Kudchokar S, Javadian MA, Dang K, Merva M, Newman M, Carrano R, Coney L, Weiner DB (1995) Induction of humoral and cellular immune responses to HIV-1 in cynomologus macaques and chimpanzees by in vivo DNA inoculation. In: Brown F, Chanock RM, Ginsberg HS, Norrby E (eds) Vaccines 95: molecular approaches to the control of infectious diseases. Cold Spring Harbor, New York, pp 99–101

Boyer J, Wang B, Ugen K, Agadjanyan MG, Javadian MA, Frost P, Dang K, Carrano R, Ciccarelli R, Coney L, Williams WV, Weiner DB (1996) In vivo protective anti-HIV immune responses in non-human primates through DNA immunization. J Med Primatol 25:242–250

Boyer J, Ugen KE, Wang B, Agadjanyan MG, Bagarazzi ML, Chattergoon M, Frost P, Javadian MA, Williams WV, Ciccarelli RB, McCallus D, Coney L, and Weiner DB (submitted) DNA vaccination as an anti HIV immunotherapy in HIV-1 infected chimpanzees J Infect Dis (in press)

Bruck C, Thiriart C, Delers A, et al. (1993a) Comparison of vaccine protection in chimpanzees immunized with two different forms of recombinant HIV-1 envelope glycoprotein. AIDS Res Hum Retroviruses 9:S110

Bruck C, Thiriart C, Fabry L, et.al. (1993b) HIV-1 envelope elicited neutralizing antibody titres correlate with protection and virus load in chimpanzees. J Cell Biochem 17:88

Bryson YJ, Pang S, Wei LS, Dickover R, Diagne A, Chen IS (1995) Clearance of HIV-1 in a perinatally infected infant. NEJM 332:833–838

Cao Y, Qin L, Zhang L, Safrit J, Ho DD (1995) Virologic and immunologic characterization of long-term survivors of human immunodeficiency virus type-1 infection. NEJM 332:201–208

Cheingsong-Papov R, Callow D, Beddows S, et al. (1992) Geographic diversity of human immunodeficiency virus type 1: serologic reactivity *env* epitopes and relationship to neutralization. J Infect Dis 165:256–261

Cheng-Mayer C, Homsy J, Evans LA, Levy JA (1988) Identification of human immunodeficiency virus subtypes and distinct patterns of sensitivity to serum neutralization. Proc Natl Acad Sci – USA 85:2815–2819

Clerici M, Shearer GM (1993) A Th1→Th2 switch is a critical step in the etiology of HIV infection. Immunol Today 14:107–111

Cocchi F, DeVico AL, Garzino-Demo A, Arya SK, Gallo RC, Lusso P (1995) Identification of RANTES, MIP-1a, MIP-1b as the major HIV-suppressive factors produced by CD_8^+ T cells. Science 270:1811–1815

Cohen J (1993) AIDS research: The mood is uncertain. Science 260:1254–1255

Coney L, Wang B, Ugen KE, Boyer JD, McCallus D, Srikantan V, Agadjanyan MG, Pachuk CJ, Herold K, Merva M, Gilbert L, Dang K, Moelling K, Newman M, Williams WV, Weiner DB (1994) Facilitated DNA inoculation induces anti-HIV-1 immunity in vivo. Vaccine 12:1545–1550

Connor R, Chen BK, Choe S, Landau NR (1995) Vpr is required for efficient replication of human immunodeficiency virus type-1 in mononuclear phagocytes. Virology 206:935–944

Cranage M, Baskerville A, Ashworth AE, et al. (1992) Intrarectal challenge of macaques vaccinated with formalin inactivated simian immunodeficiency virus. Lancet 339:273–274

Cullen B, Greene WC (1990) Functions of the auxiliary gene products of the human immunodeficiency virus type 1. Virology 178:1–5

Daniel M, Kirchoff F, Czajak SC, Sehgal PK, Desrosiers RC (1992) Protective effects of a live attenuated SIV vaccine with a deletion in the *nef* gene. Science 258:1938–1941

Danko I, Fritz JD, Shoushu J, Hogan K, Latendresse JS, Wolff JA (1994) Pharmacologic enhancement of in vivo foreign gene expression in muscle. Gene Therapy 1:1–8

Davis H, Michel ML, Whalen RG (1993) DNA-based immunization induces continuous secretion of hepatitis B surface antigen and high levels of circulating antibody. Human Mol Genetics 2:1847–1851

el-Amad Z, Murthy KK, Higgins K, Cobb EK, Haigwood NL, Levy JA (1995) Resistance of chimpanzees immunized with recombinant gp120-SF2 to challenge by HIV-1SF2. AIDS 9:1313–1322

Emini E, Schleif WA, Nunberg JH, et al. (1992) Prevention of HIV-1 infection in chimpanzees by gp120 V3 domain-specific monoclonal antibody. Nature 355:728–730

Fan L, Peden K (1992) Cell-free transmission of Vif nmutants of HIV-1. Virology 190:19–29

Fauci A (1993) Multifactorial nature of HIV disease. Implications for treatment. Science 262:1011–1015

Feng Y, Broder CC, Kennedy PE, Berger EA (1996) HIV-1 entry cofactor: functional cDNA cloning of a seven-transmembrane, G protein-coupled receptor. Science 272:872–877

Fisher A, Ensoli B, Ivanoff L, Chamberlain M, Petteway S, Rtaner L, Gallo RC, Wong-Staal F (1987) The sor gene of HIV-1 is required for efficient virus transmission in vitro. Science 237:888–895

Fultz P, Srinivasan A, Greene CR, Butler D, Swenson RB, McClure HM (1987) Superinfection of a chimpanzee with a second strain of human immunodeficiency virus. Journal of Virology 61:4026–4029

Gabuzda D, Lawrence K, Langhoff E, Terwilliger E, Dorfman T, Haseltine W, Sodroski J (1992) Role of *vif* in replication of human immunodeficiency virus type-1 transmission in CD4+ lymphocytes. J Virology 66:6489–6495

George J, Ou CY, Parekh B, et al. (1992) Prevalence of HIV-1 and HIV-2 mixed infections in Cote d'Ivoire. Lancet 340:337–339

Girard M, Kieny MP, Pinter A et al. (1991) Immunization of chimpanzees confers protection against challenge with human immunodeficiency virus. Proc Natl Acad Sci USA 88:542–546

Girard M, Meignier B, Barre-Sinoussi F, Kieny M-P, Matthews T, Muchmore E, Nara PL, Wei Q, Rimsky L, Weinhold K, Fultz PN (1995) Vaccine-induced protection of chimpanzees against infection by a heterologous Human immunodeficiency virus type-1. Journal of Virology 69:6239–6248

Golding H, Robey FA, Gates III FT, Linder W, Beining PR, Hoffman T, Golding B (1988) Identification of homologous regions in human immunodeficiency virus 1 gp41 and human MHC class II ß 1 domain. J Exp Med 167:914–923

Graham B, Wright PF (1995) Candidate AIDS vaccines. NEJM 333:1331–1339

Graham B, Keefer MC, McElrath MJ, Gorse GJ, Schwartz DH, Weinhold K, Matthews TJ, Esterlitz JR, Sinangil F, Fast PE and NIAID AVEG (1996) Safety and Immunogenicity of a candidate HIV-1 Vaccine in Healthy adults: Recombinant (RGP) 120. A randomized, double-blind trial. Ann Intern Med 125:270–279

Graziosi C, Pantaleo G, Gantt KR, Fortin JP, Demarest JF, Cohen OJ, Sekaly RP, Fauci AS (1994) Lack of evidence for the dichotomy of Th1 and Th2 predominance in HIV-infected individuals. Science 265:248–252

Greene W (1991) The molecular biology of human immunodeficiency virus type 1 infections. NEJM 324:308–317

Groux H, Torpier G, Monte D, Mouton Y, Capron A, Ameisen JC (1992) Activation-induced death by apoptosis in CD_4^+ T cells from human immunodeficiency virus-infected asymptomatic individuals. Journal of Experimental Medicine 175:331–340

Haynes J, Fuller DH, Eisenbraun MD, Ford MJ, Pertmer TM (1994) Accell particle-mediated DNA immunization elicits humoral, cytotoxic and protective responses. AIDS Res Hum Retroviruses 10:S43-45

Hilleman MR (1992) Impediments, imponderables and alternatives in the attempt to develop an effective vaccine against AIDS. Vaccine 10:1053–1058

Ho DD, Neumann AU, Perelson AS, Chen W, Leonard JM, Markowitz M (1995) Rapid turnover of plasma virions and CD_4^+ lymphocytes in HIV-1 infection. Nature 373:123–126

Hulskotte EG, Geretti A-M, Siebelink KH, van Amerongen G, Cranage MP, Rud EW, Norley SG, de Vries P, Osterhaus AD (1995) Vaccine-induced virus neutralizing antibodies and cytotoxic T cells do not protect macaques from experimental infection with Simian immunodeficiency virus $SIV_{mac32H(J5)}$. Journal of Virology 69:6289–6296

Johnson P, Montefiori DC, Goldstein S, Hamm TE, Zhou J, Kitov S, Haigwood NL, Misher L, London WT, Gerin JL et al. (1992) Inactivated whole virus vaccine derived from a proviral clone of simianimmunodeficiency virus induces high levels of neutralizing antibodies and confers protection against heterologous challenge. Proc Natl Acad Sci USA 89:2175–2179

Jowett J, Planelles V, Poon B, Shah NP, Chen ML, Chen ISY (1995) The human immunodeficiency virus type 1 *vpr* gene arrests infected T cells in the G2 + M phase of the cell cycle. J Virol 69:6304–6313

Katzenstein D, Vujcic LK, Latif A, Boulos R, Halsey NA, Quinn TC, et al. (1990) Human immunodeficiency neutralizing antibodies in sera from North Americans and Africans emergence of neutralization. J Acquir Immune Syndr 3:810–816

Kestler H, Ringler DJ, Mori K, Panicalli DL, Sehgai PK, Daniel MD, Desrosiers RC (1991) Importance of the *nef* gene for maintenance of high virus loads and for the development of AIDS. Cell 65:651–662

Kim J, Chattergoon M, Ayyavoo V, Bagarazzi ML, Boyer JD, Wang B, Weiner DB (1997a) Development of a multi-component candidate vaccine for HIV-1. Vaccine 15:879–883

Kim J, Bagarazzi ML, Hu Y, Trivedi N, Chattergoon MA, Dang K, Agadjanyan MG, Mahalingam S, Boyer JD, Wang B, Weiner DB (1997b) Engineering of in vivo immune responses to DNA immunization via co-delivery of costimulatory molecule genes. Nature Biotech (in press)

Kim J, Ayyavoo V, Bagarazzi ML, Chattergoon MA, Dang K, Wang B, Boyer JD, Weiner DB (1997c) In vivo engineering of a cellular immune response by co-administration of IL-12 expression vector with a DNA immunogen. J Immunol 158:816–826

Kim S, Ikeuchi K, Byrn R, Groopman J, Baltimore D (1989) Lack of a negative influence on viral growth by the *nef* gene of human immunodeficiency virus type 1. Proc Natl Acad Sci USA 86:9544–9548

Kirchoff F, Greenough TC, Brettler DB, Sullivan JL, Desrosiers RC (1995) Brief report: absence of intact *nef* sequences in a long-term survivor with nonprogressive HIV-1 infection. NEJM 332:228–232

Klimkait T, Strebel K, Hoggan MD, Martin MA, Orenstein JM (1990) The human immunodeficiency virus type-1 specific protein Vpu is required for efficient virus mutation and release. J Virol 64:621–629

Koup R, Safrit JT, Cao Y, et al. (1994) Temporal association of cellular immune responses with the initial control of viremia in primary Human immunodeficiency virus type-1 syndrome. J Virol 68:4650–4655

Learmont J, Tindall B, Evans L, Cunningham A, Cunningham P, Wells J, Penny R, Kaldor J, Cooper DA (1992) Long-term symptomless HIV-1 infection in recipients of blood products from a single donor. Lancet 340:863–867

Levine B, Mosca JD, Riley JL, Carroll RG, Vahey MT, Jagodzinski LL, Wagner KF, Mayers DL, Burke DS, Weislow OS, St. Louis DC, June CH (1996) Antiviral effect and ex vivo CD4+ T cell proliferation in HIV-positive patients as a result of CD28 costimulation. Science 272:1939–1943

Levy D, Refaeli Y, Weiner DB (1995) The *vpr* regulatory gene of human immunodeficiency virus. In: Chen I, Koprowski H, Srinivasan A, Vogt PK (eds) Transacting Functions of Human Retroviruses. Springer, Berlin, Heidelberg, New York 193:209–236

Levy DN, Fernandes LS, Williams WV and Weiner DB (1993) Induction of cell differentiation by human immunodeficiency virus 1vpr. Cell 72:541–550

Li J, Lord CI, Haseltine W, Letvin NL, Sodroski J (1992) Infection of cynomolgus monkeys with a chimeric HIV-1/SIV_{mac} virus that expresses the HIV-1 envelope glycoproteins. J Acquir Immune Defic Syndr 5:639–646

Lu S, Santoro JC, Fuller DH, Haynes JR, Robinson HL (1995) Use of DNAs expressing HIV-1 *env* and non-infectious HIV-1 particles to raise antibody responses in mice. Virology 209:147–154

Lu S, Arthos J, Montefiori DC, Yasutomi Y, Manson K, Mustafa F, Johnson E, Santoro JC, Wissink J, Mullins JI, Haynes JR, Letvin NL, Wyand M, Robinson HL (1996) Simian immunodeficiency virus DNA vaccine trial in macaques. J Virol 70:3978–3991

MacGregor R, Gluckman S, Lacy K, Wang B, Ugen K, Chattergoon M, Bagarazzi ML, Williams W, Ginsberg R, Higgins T, Boyer JD, Weiner DB (1996a) A DNA plasmid vaccine for HIV-1: experience in the first human trial indicates humoral and cell-immune responses. Fourth Conference on Retroviruses and Opportunistic Infections, Washington DC

MacGregor R, Gluckman S, Lacy K, Kaniefski B, Boyer JD, Wang B, Bagarazzi ML, Williams WV, Francher D, Ginsberg R, Higgins T, Weiner DBU (1996b) First human trial of a facilitated DNA plasmid vaccine for HIV-1: safety and host response. XI International Conference on AIDS, Vancouver

Mackewicz CE, Ortega H, Levy JA (1994) Effect of cytokines on HIV replication in CD_4^+ lymphocytes: lack of identity with the CD_8^+ cell antiviral factor. Cell Immunol 153:329–343

Malim M, Hauber J, Le SY, Maizel JV, Cullen BR (1989) The HIV-1 *rev trans*-activator acts through a structured target sequence to activate a nuclear export of unspliced viral mRNA. Nature 338:254–257

Michael N, Chang G, D'Arcy LA, Ehrenberg PK, Mariani R, Busch MP, Birx DL, Schwartz DH (1995) Defective accessory genes in a human immunodeficiency vitus type-1 infected long-term survivor lacking recoverable virus. J Virol 69:4228–4236

Musgrove P (1993) Investing in health. the 1993 world development report of the World Bank. Bulletin of the Pan-American Health Organization 27:284–286

Narayan O, Joag SV, Stephens EB (1995) Selected models of HIV-induced neurological disease. Current topics in microbiology and immunology, Springer, Berlin, Heidelberg, New York 202:151–166

Nixon D, Townsend AR, Elvin JG, Rizza CR, Gallwey J, McMichael AJ (1988) HIV-1 gag-specific cytotoxic T lymphocytes defined with recombinant vaccinia virus and synthetic peptides. Nature 336:484–487

Novembre F, O'Neil SP, Saucier M, Anderson DC, Klumpp SA, Brown CR, Hart CA, Guenther PC, Swenson RB, McClure HM (1996) Pathogenic HIV-1 infection in chimpanzees. Fourteenth annual symposium on nonhuman primate models for AIDS. Portland, OR

Oberlin E, Amara A, Bacherlerie F, Bessia C, Virelizier J-L, Arenzana-Seisdedos F, Schwartz O, Heard J-M, Clark-Lewis I, Legler DF, Loetscher M, Baggiolini M, Moser B (1996) The CXC chemokine SDF-1 is the ligand for LESTR/fusin and prevents infection by T-cell-line-adapted HIV-1. Nature 382:833–835

Okuda K, Bukawa H, Hamajima K, Kawamoto S, Sekigawa K-I, Yamada Y, Tanaka S-I, Ishii N, Aoki I, Nakamura M, Yamamoto H, Cullen BR, Fukushima J (1995) Induction of potent humoral and cell-mediated immune responses following direct injection of DNA encoding the HIV type 1 *env* and *rev* gene products. AIDS Res Hum Retroviruses 11:933–943

Pantaleo G, Menzo S, Vaccarezza M, Graziosi C, Cohen OY, Demarest JF, Montefiori D, Orenstein JM, Fox C, Schrager LK, et al. (1995) Studies in subjects with long-term nonprogressive human immunodeficiency virus infection. NEJM 332:209–216

Pauza C, Emau P, Salvato MS, et al. (1993) Pathogenesis of SIV_{mac251} after atraumatic inoculation of the rectal mucosa in Rhesus monkeys. Journal of Medical Primatology 22:154–161

Peeters M, Gershy-Dame GM, Fransen K, et al. (1992) Virological and polymerase chain reaction studies of HIV-1/HIV-2 dual infection in Cote d'Ivoire. Lancet 340:339–340

Phillips R, Rowland-Jones S, Nixon DF, Gotch FM, Edwards JP, Ogunlesi AO, et al. (1991) Human immunodeficiency virus genetic variation that can escape cytotoxic T cell recognition. Nature 354:453–459

Pialoux G, Excler JL, Riviere Y, Gonzalez-Canali G, Feuillie V, Coulaud P, Gluckman JC, Matthews TJ, Meignier B, Kieny MP, et al. (1995) A prime-boost approach to HIV preventive vaccine using a recombinant canarypox virus expressing glycoprotein 160 (MN) followed by a recombinant gp160 (MN/LAI). AIDS Res and Human Retrovir 11:373–381

Pinto L, Sullivan J, Berzofsky JA, Clerici M, Kessler HA, Landay AL, Shearer GM (1995) Env-specific cytotoxic T lymphocyte responses in HIV seronegative health care workers occupationally exposed to HIV-contaminated body fluids. J Clin Invest 96:867–876

Putkonen P, Thorstensson R, Ghavamzadeh L, Albert J, Hild K, Biberfeld G, Norrby E (1991) Prevention of HIV-2 and SIV_{SM} infection by passive immunization in cynomolgus monkeys. Nature 352:436–438

Refaeli Y, Levy DN, Weiner DB (1995) The glucocorticoid receptor type II complex is a target of the HIV-1 *vpr* gene product. Proc Natl Acad Sci USA 92:3621–3625

Riviere Y, Tannequ-Salvadori F, Regnault A, Lopez A, Sansonetti P, Guy B, et al. (1989) Human immunodeficiency virus-specific cytotoxic responses of seropositive individuals: distinct types of effector cells mediate killing of targets expressing Gag and Env proteins. Journal of Virology 63:2270–2277

Rogel M, Wu LI, Emerman M (1995) The human immunodeficiency virus type 1 *vpr* gene prevents cell proliferation during chronic infection. J Virol 69:882–888

Rosen C (1991) Regulation of HIV gene expression by RNA-protein interaction. Trends Gen 7:9–14

Rosenthal K (1993) Recombinant adenoviruses:A vector for all seasons. AIDS Res Hum Retroviruses 9:S28

Rowland-Jones S, Nixon DF, Aldhous MC (1993) HIV-specific cytotoxic T-cell activity in an HIV-exposed but uninfected infant. Lancet 341:860–861

Rowland-Jones S, Sutton J, Ariyoshi K, Dong T, Gotch F, McAdam S, Whitby D, Sabally S, Allimore A, Corrah T, Takiguchi M, McMichael A, Whittle H (1995) HIV-specific T-cells in HIV-exposed but uninfected Gambian women. Nature Medicine 1:59–64

Saag M, Hahn BH, Gibbons JH, Li Y, Parks ES, Parks WP, Shaw GM (1988) Extensive variation of human immunodeficiency virus type-1 in vivo. Nature 334:440–444

Schultz A, Hu S (1993) Primate models for HIV vaccines. AIDS 7:5161–5170

Schwartz D, Gorse G, Clements ML, Belshe R, Izu A, Duliege A-M, Berman P, Twaddell T, Stablein D, Sposto R, Siliciano R, Matthews T (1993) Induction of HIV-1 neutralising and syncytium-inhibiting antibodies in uninfected recipients of HIV-1(IIIB) rpg120 subunit vaccine. Lancet 342:69–73

Sheppard H, Bridges SH, Mathieson BJ, Walker MC, Weinhold K (1994) Conference on advances in AIDS vaccine development – 1993 Summary: correlates of HIV Immunity Working Group. AIDS Res Hum Retroviruses 10:S171–176

Shibata R, Hoggan MD, Broscius C, Englund G, Theodore T, Buckler-White A, Arthur L, Israel Z, Schult A, Lane HC, Martin MA (1995) Isolation and characterization of a syncytium-inducing, macrophage/T-cell line tropic HIV-1 isolate that readily infects chimpanzee cells in vitro and in vivo. Journal of Virology 69:4453–4462

Tang D, Devit M, Johnston SA (1992) Genetic immunization is a simple method for eliciting an immune response. Nature 356:152–154

Taylor R (1992) Does HIV-1 kill lymphocytes by inducing cell suicide? J NIH Res 4:65–69

Thomason D, Booth FW (1990) Stable incorporation of a bacterial gene into adult rat skeletal muscle in vivo. Am J Cell Physiol 258:C578–581

Ugen K, Goedert JJ, Boyer JD, Refaeli Y, Frank I, Williams WV, Willoughby A, Landesman S, Mendez H, Rubinstein A, Keiber-Emmons T, Weiner DB (1992) Vertical transmission of human immunodeficiency virus (HIV) infection. Reactivity of maternal sera with glycoprotein 120 and 41 peptides from HIV type 1. J Clin Invest 89:1923–1930

Ugen K, Boyer JD, Wang B, Gluckman S, Williams WV, Nyland S, Carmack B, Agadjanyan MG, Bagarazzi ML, Higgins T, Coney L, Ciccarelli R, Ginsberg R, MacGregor RR, Weiner DB (1997a) Facilitated nucleic acid vaccination of humans with an HIV-1 expressing DNA boosts immune responses against the envelope glycoprotein. In: Brown F, Chanock RM, Ginsberg HS, Norrby E (eds) Vaccines 97: molecular approaches to the control of infectious diseases. Cold Spring Harbor, pp 145–150

Ugen K, Boyer JD, Wang B, Bagarazzi ML, Javadian MA, Frost P, Merva MM, Nyland S, Williams WV, Coney L, Ciccarelli R, Weiner DB (1997b) Nucleic acid immunization of chimpanzees as a prophylactic/immunotherapeutic vaccination model for HIV-1: prelude to a clinical trial. Vaccine 15:927–930

Ulmer J, Donnelly J, Parker SE, Rhodes GH, Felgner PL, Dwarki VL, et al. (1993) Heterologous protection against influenza by injection of DNA encoding a viral protein. Science 259:1745–1749

von Schwedler U, Song J, Aiken C, Trono D (1993) Vif is crucial for human immunodeficiency virus type 1 proviral DNA synthesis in infected cells. J Virol 67:4945–4955

Walker B, Flexner C, Paradis TJ, Fuller TC, Hirsch MS, Schooley RT, Moss B (1988) HIV-1 reverse transcriptase is a target for cytotoxic T lymphocytes in infected individuals. Science 240:64–66

Walker BD (1994) The rationale for immunotherapy in HIV-1 infection. J Acquir Immune Defic Syndr 7:S6–13

Wang B, Boyer JD, Srikantan V, Coney L, Carrano R, Phan C, Merva M, Dang K, Agadjanyan MG, Ugen KE, Williams WV, Weiner DB (1993a) DNA inoculation induces neutralizing immune responses against Human immunodeficiency virus type 1 in mice and non-human primates. DNA Cell Biol 12:799–805

Wang B, Ugen KE, Srikantan V, Agadjanyan MG, Dang K, Refaeli Y, Sato AI, Boyer J, Williams WV, Weiner DB (1993b) Gene inoculation generates immune responses against human immunodeficiency virus type 1. Proc Natl Acad Sci USA 90:4156–4160

Wang B, Boyer JD, Ugen KE, Merva M, Dang K, Agadjanyan MG, Williams WV, Weiner DB (1994a) Analysis of antigen-specific immune responses induced through in vivo genetic inoculation. In: Brown F, Chanock RM, Ginsberg HS, Norrby E (eds) Vaccines 94: molecular approaches to the control of infectious diseases. Cold Spring Harbor, New York, pp 83–90

Wang B, Merva M, Dang K, Ugen KE, Boyer JD, Williams WV, Weiner DB (1994b) DNA inoculation induces protective in vivo immune responses against cellular challenge with HIV-1 antigen-expressing cells. AIDS Res Hum Retroviruses 10:S35–41

Wang B, Merva M, Dang K, Ugen KE, Williams WV, Weiner DB (1995a) Immunization by direct DNA inoculation induces rejection of tumor cell challenge. Human Gene Ther 6:407–418

Wang B, Boyer JD, Srikantan V, Ugen KE, Gilbert L, Phan C, Dang K, Merva M, Agadjanyan MG, Newman M, Carrano R, McCallus D, Coney L, Williams WV, Weiner DB (1995b) Induction of humoral and cellular immune responses to the human immunodeficiency type-1 virus in non-human primates by in vivo DNA inoculation. Virology 211:102–112

Wang B, Boyer JD, Ugen KE, Srikantan V, Ayyavoo V, Agadjanyan MG, Williams WV, Newman M, Coney L, Carrano R, Weiner DB (1995c) Nucleic acid-based immunization against HIV-1: induction of protective in vivo immune responses. AIDS 9:S159–170

Wang B, Ge YC, Palasanthiran P, Xiang S-H, Ziegler J, Dwyer DE, Randle C, Dowton D, Cunningham A, Saksena NK (1996) Gene defects clustered at the C-Terminus of the vpr gene of HIV-1 in long-term nonprogressing mother and child pair: In vivo evolution of vpr quasispecies in blood and plasma. Virology 223:224–232

Wei X, Ghosh SK, Taylor ME, Johnson VA, Emini EA, Deutsch P, Lifson JD, Bonhoeffer S, Nowak MA, Hahn BH, Saag MS, Shaw GM (1995) Viral dynamics in human immunodeficiency virus type-1 infection. Nature 373:117–122

Yasutomi Y, Robinson HL, Lu S, Mustafa F, Lekutis C, Arthros J, Mullins JI, Voss G, Manson K, Wyand M, Letvin NL (1996) Simian immunodeficiency virus-specific cytotoxic T-lymphocyte induction through DNA vaccination of rhesus monkeys. J Virol 70:678–681

Human Immunodeficiency Virus Immunotherapy Using a Retroviral Vector

J.F. WARNER[1,2], D.J. JOLLY[1], and J. MERRITT[1]

1 Introduction

Gene transfer technology is an effective means of introducing genes and ultimately providing immunogenic proteins for activation of the immune response. Conventional immunization procedures are capable of providing proteins to the exogenous antigen presentation pathway for activation of $CD4^+$ T cells and eventual antibody production. However, intracellular synthesis of foreign proteins appears to favor optimal antigen processing/presentation events involved in the consistent activation of $CD8^+$ cytotoxic T lymphocytes (CTLs) (BRACIALE et al. 1987; GERMAIN and MARGULIES 1993) which recognize antigen in the context of class I major histocompatibility complex (MHC) molecules. Gene transfer systems, thus, may provide a consistent way of delivering protein antigens to the endogenous antigen presentation pathway for activation of cellular immunity, particularly CTL activation.

Retroviral vectors can efficiently deliver genes to mammalian cells. The retroviral vector is a molecularly engineered, nonreplicating gene delivery system with the capacity to carry approximately 8 kilobases (kb) of genetic information for a

[1]Chiron Viagene, 11055 Roselle Street, San Diego, CA 92121, USA
[2]Current address: Inex Pharmaceuticals Corp., 1779 West 75th Ave., Vancouver, BC, Canada V6P6P2

variety of antigens and can potentially be generated with tropisms for infecting various types of cells from different species. Retroviral vector transduction involves a stable gene integration event without detectable cytopathicity and is carried out in the absence of any associated viral gene expression. In this manner, the retroviral vector potentially provides a mechanism for repeated delivery of only a particular gene (i.e., antigen) of interest for immune activation.

In many viral infections the cellular immune response plays an important role in eliminating virus-infected cells. CTLs specifically kill virus infected cells and are important in limiting the severity and duration of many viral diseases (Kast et al. 1986; Klavinskis et al. 1989; Rouse et al. 1988; Reddehasse et al. 1987; Reusser et al. 1991). In this context, CTLs may play a role in controlling early disease progression in individuals infected with the human immunodefiency virus (HIV-1) believed to be the cause of acquired immunodeficiency syndrome (AIDS). HIV-infected individuals have been shown to possess CTLs specific for various HIV proteins (Hoffenbach et al. 1989; Carmichael et al. 1993). A temporal association between increased viral burden (Ho et al. 1995; Wei et al. 1995) and decreased CTL activity has been observed in some HIV-seropositive individuals (Koup et al. 1994; Riviere et al. 1989; Schrager et al. 1994).

Our initial studies employing retroviral vector-mediated gene transfer involved the administration of ex vivo vector-transduced syngeneic and autologous cells to experimental animals (Warner et al. 1991) and humans (Ziegner et al. 1995), respectively. The retroviral vector (HIV env/rev) employed in these studies encoded the HIV-1 IIIB envelope (env) and rev proteins as well as a neomycin-phosphoryltransferase resistance (neo^r) selectable marker. Tissue biopsies were obtained which provided cells that were transduced in vitro with the vector, the cells were selected (neo^r) for gene expression, and the gene-modified cells infused into mice and rhesus macaques (Laube et al. 1994). These studies demonstrated the ability of retroviral vector-transduced cells to induce HIV-1env/rev-specific $CD8^+$, MHC-restricted CTLs and antibody responses in murine and nonhuman primate models. Recently, these studies have been extended to humans, in which $CD8^+$, class I MHC-restricted CTL activity specific for HIV-1env/rev appeared to be augmented in HIV-seropositive, asymptomatic subjects following administration of retroviral vector-transduced autologous fibroblasts (Ziegner et al. 1995). No adverse clinical events were observed in these patients. These studies, hence, provided the initial basis for safety and biological activity of an HIV immunotherapeutic application of gene transfer technology using retroviral vectors.

The ex vivo gene transfer strategy for augmenting CTL activity in HIV-infected individuals, however, was deemed less practical in this case due to the extensive cell culturing requirements which would limit treatment to large numbers of infected individuals, thus, warranting the development of an alternative means of gene transfer. The direct, in vivo administration of the retrovial vector provided a more feasible approach for gene therapy treatment. The following discussion describes the development of the direct vector administration approach in murine, rhesus monkey, and baboon animal models using the HIV env/rev retroviral vector. Based on the results obtained from these animal studies, human clinical trials have been initiated involving the use of the HIV env/rev retroviral vector in HIV-1 seropositive, asymptomatic subjects.

2 Retroviral Vector

A nonreplicating, recombinant retroviral vector encoding the env and rev proteins from the HIV-1 IIIB viral strain (HIV env/rev) was employed to provide endogenous production of these HIV-1 proteins in murine, nonhuman primate, and human cells. The N2 IIIBenv/rev retroviral vector construct was generated by inserting the 3.1 kb *XhoI-ClaI* fragment from the pAFenv retroviral provector containing the HIV-1IIIB *env* and *rev* genes (WARNER et al. 1991) into the backbone of an engineered N2 murine recombinant retrovirus. For this N2 retroviral construct, the ATG initiator codon for Moloney murine leukemia virus (MLV) *gag* was converted to ATT by site-directed mutagenesis, resulting in an increased level of expression of transduced genes. For selection purposes, the neo^r gene, conferring resistance to the antibiotic geneticin (G418, GIBCO, Grand Island, NY), was included in the vector (CHADA et al. 1993). The N2 IIIBrev vector, which encodes the HIV- I rev protein and a neo^r marker, was generated by inserting the rev cDNA coding sequence from pcREV (MALIM et al. 1988) into the N2 backbone at *XhoI-ClaI* restriction sites. In addition, the bacterial *lacz* gene encoding β-galactosidase (βgal) (PRICE et al. 1987) was cloned into the N2 backbone as described above to generate the N2 βgal vector.

Novel retroviral packaging cell lines have been developed for these studies based upon the expression of MLV *gag/pol* and *env* sequences in the canine D-17 cell line (ATCC #CCL 183). These packaging cell lines were developed using the principle of splitting the retroviral genome, removing the viral long terminal repeat (LTR) sequences and replacing them with the cytomegalovirus (CMV) immediate early promoter (BOSHART et al. 1985) to reduce the potential of generating replication competent retrovirus (Packaging Cells Patent 1991, PCT Application 91/06852). Producer cell lines were generated by transduction of vesicular stomatitis virus (VSV)-G pseudotyped vector (BURNS et al. 1993) into the appropriate packaging cell lines.

DA/N2 IIIBcnv producer cells were grown and the supernatant containing replication defective HIVenv/rev vector was collected, filtered (0.45 μ), concentrated by ultrafiltration and column purified. The purified preparation of vector was stored either at −80 °C as a frozen liquid or at −20 °C in lyophilized form until use. Both frozen and lyophilized vector preparations appear to have similar biological activity when tested at equivalent vector titers. The titer of vector preparations was determined by serial dilution and colony forming unit assay on HT1080 indicator cells selected in G418 containing medium. Titers were typically between 10^6 and 10^7 cfu/ml.

Aliquots of vector (0.2 ml) were tested for the presence of replication competent retrovirus (RCR) by the extended S^+/L^- assay (PEEBLES et al. 1975) and marker rescue assay (Printz et al. 1994). The vector producing cell line was also tested by cocultivation with *Mus dunni* cells (Lander and Chattopadhyay 1984) to confirm the absence of RCR in the producer cells.

3 Animal Studies

3.1 Immune Response Induction in Mice

We have evaluated the ability of the HIV env/rev retroviral vector to induce CTLs and antibody reactivity in a murine model. The generation of HIV-1-specific CTL responses directed against the gene products (HIV-1 env, HIV-1 rev and neo^r) delivered by the HIV env/rev vector was evaluated in BALB/c mice. Mice were injected by the intramuscular (IM) route in both legs with 100 µl of 10^6 cfu/ml vector. Splenocytes were obtained from vector immunized mice 7 days later, stimulated in vitro for 5–7 days with a syngeneic fibroblast cell line [BC10ME (BC)] (Patek et al. 1982) transduced with the HIV env/rev vector and expressing the HIV-1 env and rev proteins and the neo^r marker (BC-env/rev). The resulting effector cells were tested for cytolytic activity on BCenv/rev, BCβgal, and BC target cells using a standard ^{51}Cr-release cytotoxicity assay. HIV-1 env specific cytotoxic activity was demonstrated on BCenv/rev and BCrev target cells, but not control BCβgal or BC target cells (Fig. 1). The results suggested the cytolytic activity was specific for HIVenv/rev determinants and not neo^r determinants. The cytolytic activity was shown to be mediated by $CD8^+$ CTLs using T cell-specific monoclonal antibody depletion. The HIV-specific murine CTLs showed lysis on syngeneic $H\text{-}2^d$ targets, but did not lyse allogeneic target cells expressing HIV-1 env and rev proteins (BL6-env/rev, $H\text{-}2^b$), indicating that the mouse CTLs were MHC-restricted. Related studies using monoclonal antibody directed against different H-2 alleles indicated that CTLs induced with vector-transduced cells expressing HIV-1IIIBenv/rev proteins is primarily restricted by the $H\text{-}2D^d$ allele (Warner et al. 1991). Vector inactivation studies showed that CTL induction was a function of active gene expression and not the result of soluble or particulate HIV gpl6O/l2O antigen potentially present in the vector preparation. Therefore, conventional class I MHC-restricted, CD8+ CTLs were generated in mice as a result of the direct in vivo administration of the HIVenv/rev vector.

Examination of CTL induction in BALB/c mice with respect to vector dose showed a threshold relationship. By injecting mice with serial dilutions of the HIVenv/rev vector ranging from 10^5 cfu to 10^7 cfu, splenocyte effectors showed equivalent target cell lysis. Mice injected with a lower quantity of vector (10^3 cfu or 10^4 cfu) exhibited negligible target cell lysis. The frequency of responding mice increased with higher vector doses (i.e., 10^6–10^7 cfu), in particular, doses of $<10^5$ cfu often times resulted in CTL responses from 40% of the mice, whereas doses of $>10^6$ cfu typically demonstrated 90%–100% responding mice. Analysis of

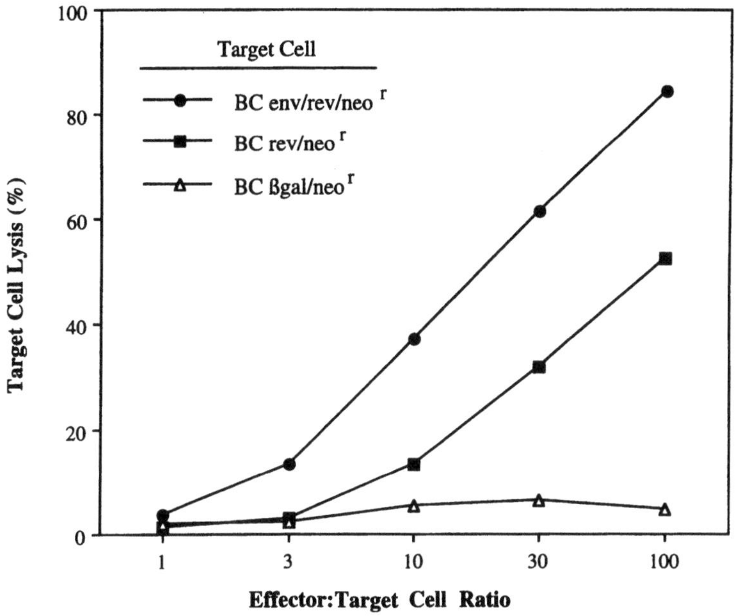

Fig. 1 HIVenv/rev retroviral vector induced cytotoxic T lymphocyte (CTL) activity in mice. Splenocytes obtained from BALB/c mice injected intramuscularly with 10^6 cfu of vector were stimulated in vitro with HIV-1IIIBenv/rev expressing cells. The resulting effectors were tested at various effector:target cell ratios on syngeneic fibroblast target cells expressing HIV-1 IIIB env, rev and neo^r proteins (BCenv/rev), HIV-1IIIB rev and neo^r (BCrev), and control targets expressing β-galactosidase and neo^r (BCβgal) in a standard ^{51}Cr-release assay

these data suggested that in mice, approximately 10^6 cfu of vector were necessary for reproducible CTL induction.

Because the HIVenv/rev vector encodes different proteins (env, rev, neo^r), vector-induced CTLs were tested on various target cells to evaluate the specificity of the vector-induced CTL response. HIVenv/rev vector-induced CTL cell lines recognized and lysed BC target cells infected with the vaccinia-gp16O recombinant virus which does not express the rev protein, indicating that HIV-1 env-specific CTL were present in the nonclonal effector cell population. These CTLs were also evaluated for rev-specific effector cells in which target cells transduced with a retroviral vector encoding only HIV-1 rev (BCrev) were employed. A CTL cell line showed rev-specific target cell lysis, as BCrev targets and BCenv/rev targets were lysed, but not control BCβgal targets. To further characterize the env-specific component of the CTL response, a peptide corresponding to the HIV-1 IIIB hypervariable V3 loop region of the env protein was used to pulse-label target cells. This peptide is an immunodominant CTL epitope [amino acids 315–329; RIQRGPGRAFVTIGK(C)], recognized by BALB/c T cells specific for the HIV-1 IIIB env protein (Takahashi et al. 1988). Target cells pulsed with this peptide were recognized and lysed by HIV env/rev induced CTLs in contrast to BC control targets. Direct HIV env/rev vector administration to mice also induced a cross-reactive CTL response as demonstrated by lysis of targets coated with a peptide derived from the equivalent V3 hypervariable loop region [amino acids 315–329; RIHIGPGRAFYTTKN(C)] of the HIV-1MN viral isolate (M. Irwin, unpublished data). These results suggested that the vector-induced CTL population was capable of recognizing HIV-1 IIIB env and rev determinants as well as determinants cross-reactive with an env protein from a different HIV-1 isolate.

Additional studies have evaluated the ability of HIV env/rev vector-induced CTLs to recognize and lyse HIV-infected target cells. We have constructed a human cell line (Hu/D^d) expressing both the CD4 receptor (HIV receptor) (Maddon et al. 1986) and the murine H-$2D^d$ MHC molecule (BALB/c mouse T cell restriction element). Unlike murine cells, the Hu/D^d cell line should be infectable by HIV-1 viral strains and could also function as a target for murine CD8+, class I MHC-restricted (i.e., H-$2D^d$) CTLs directed against HIV-1IIIB viral determinants. Hu/D^d cells were infected with various HIV viral strains, and approximately 10 days post-infection the infected cells were used as targets in a ^{51}Cr release assay using murine CTLenv/rev obtained from BALB/c mice immunized with BCenv/rev transduced cells. CTLenv/rev exhibited substantial lysis of target cells infected with the HIV-1 IIIB virus strain. Hu/D^d target cells infected with HIV-1 MN, WMJ II, SF2 and CC viral isolates were also lysed by CTLenv/rev to a lesser extent (i.e., 50%) than HIV-1 IIIB infected targets (Chada et al. 1993). Target cells infected with HIV-2 (CBL20) were not lysed by CTLenv/rev. Lysis of HIV-1 infected Hu/D^d targets was shown to be dependent on CD8+ T cells. These results indicated that antigenic CTL determinants presented by vector-transduced cells were similar to those expressed following viral infection and that those determinants could be presented in the context of the murine H-$2D^d$ MHC molecule.

Target cells infected with clinical strains obtained from three patients also exhibited substantial lysis by CTLenv/rev, although the level of lysis was lower than that observed for HIV-1 IIIB infected cells, but similar to the level of lysis observed on HIV-1 MN virus-infected targets. The pattern of CTL cross-reactivity was consistent over a period of several weeks after HIV-1 infection of the Hu/D^d cells. The Hu/D^d target cells, infected with the divergent HIV isolates, were p24 antigen positive and exhibited substantial levels of env expression by western blot analysis. Sequence analyses of V3 gp120 envelope regions and neutralizing antibody assays have indicated that MN-type viruses represent 60%–70% of clinical HIV-1 isolates in the North American patient population, while HIV-1IIIB-type viruses represent only 7%–14% (LaRosa et al. 1990). The HIV-1 IIIB and MN strains share approximately 80% sequence similarity in their envelope sequences. Therefore, CTLenv/rev were able to recognize and specifically lyse targets infected with a number of divergent prototypic and clinical HIV-1 strains suggesting broader cross-reactivity at the level of CTL activity than antibody recognition. These data suggest that the envelope and/or rev proteins from divergent HIV-1 isolates, including patient isolates, may share common epitope(s) that can be recognized by the murine CTLenv/rev population.

As another assay of the biological function of CTLs, we determined whether CTLenv/rev could diminish HIV-1 replication in infected cultures. HIV-1IIIB infected Hu/D^d cells were mixed with CTLenv/rev effectors at varying ratios and the culture incubated for 2 weeks. A reduction in HIV-1 IIIB titer of greater than three logs was observed using a syncytium assay following the addition of the highest number of CTLenv/rev effectors (i.e., 100:1, E:T). At lower input levels of CTLenv/rev cells, reduced levels of inhibition of HIV-1 replication were found, although at an E:T ratio of 10:1, a 50-fold reduction in HIV-1 titer was still observed, indicating the ability of these murine CTLenv/rev to inhibit HIV-1IIIB growth in tissue culture. Inhibition of HIV-1 replication by CTLenv/rev was also observed in cells infected with HIV-1 MN virus (S. Chada, unpublished data).

HIV-1 env-specific antibody responses were also examined from mice injected weekly with 10^6 cfu of the HIVenv/rev retroviral vector. Serum samples obtained from animals prior to and after injection of HIVenv/rev were tested by ELISA for HIV-1 gp120-specific antibody. Moderate antibody (IgG) responses were observed in BALB/c mice after the second administration of 10^6 cfu of vector and persisted for greater than 6 months following the third injection. IgM HIV-1 gpl20-specific antibody was present following the first vector injection, whereas, pre-injection sera showed no anti-gp120-specific antibody. Injection doses of 10^7 cfu vector elicited increased anti-gp120-specific antibody activity. Therefore, HIV env/rev vector immunization appears capable of activating humoral immune responses in the murine animal model.

3.2 In Vivo Vector Expression in Mice

We have also initiated studies to examine the fate of direct IM administration of retroviral vector in the murine model. BALB/c mice were injected in the gastrocnemius muscle with purified, formulated retroviral vector preparations to evaluate: (a) vector trafficking in males and females; (b) in vivo gene expression; and (c) which cell types infiltrate the injection site following IM injection.

The HIVenv/rev retroviral vector was administered IM in 40 male mice and the animals were killed to evaluate various organs after time periods ranging from 2 h to 21 days. Genomic DNA from the injection site muscle, liver, spleen, kidney, thymus and testes was isolated and assayed by polymerase chain reaction (PCR) analysis for the presence of provector DNA. All animals injected with vector were positive for provector sequences at the injection site (Sajjadi et al. 1994). The injection site muscles were found to be positive by PCR for at least 21 days. The testes, liver, spleen and kidney were negative in all animals. All samples from 32 animals injected with diluent control were negative by PCR. We have also examined 20 female mice for the presence of provector sequences after IM administration of HIVenv/rev retroviral vector. The female mice were injected using the same regimen as used in the male mice; however, the set of tissues examined was expanded to include ovaries, brain, lung and lymph nodes. In these assays, the major positive PCR signals were obtained from injection site muscles and draining inguinal lymph nodes. The presence of provector sequences in the lymph node may suggest that the vector can directly traffic to the draining lymph node or that vector-transduced cells are capable of migrating into this site. The ovaries of the 20 treated female mice did not show, at any timepoint, the presence of provector sequences, with the exception of a positive PCR signal (at the 2 h timepoint) in one ovary. Thus, these results indicate that the vector gene sequences do appear to be integrated at the site of injection and to a lesser extent in lymph node tissue with no significant involvement of germinal tissue (i.e., testes or ovaries).

Retroviral vectors encoding the bacterial βgal gene were employed to localize protein expression in individual cells. The vector was injected IM and shown to induce CTL responses against βgal-expressing syngeneic target cells (BCβgal). Analysis of the injection site demonstrated βgal expression in individual myofibers. Retroviral vectors typically transduce dividing cells, therefore, it was surprising that βgal-expressing myofibers were consistently observed because myofibers are terminally differentiated cell syncytia generated by fusion of individual myoblasts. These data, however, are consistent with the hypothesis that retroviral vectors transduce myoblasts in vivo and the resulting transduced cells fuse with the existing muscle cells to form myofibers expressing the delivered gene product.

Injection site muscles were also examined by immunohistochemistry to define infiltrating cells. Substantial infiltration of MHC class II positive cells, including macrophages and B cells, and CD4+, but not CD8+, T cells was observed. However, control experiments using injection of diluent formulation buffer demonstrated similar, but reduced, cell infiltration. These data indicate that IM injection of retroviral vectors does not cause gross tissue destruction (myopathy) or an increase in inflammatory cells beyond that caused by a needle injury. The results suggest that the CD8+ T cell activation occurs distal to the injection site, likely in the draining lymph nodes. Future studies will need to define the cell type(s) involved in actual antigen presentation to CTL precursors.

3.3 Immune Response Induction in Nonhuman Primates

Retroviral vector immunization of nonhuman primates was performed to evaluate immune response induction and safety. Induction of cytotoxic activity by direct injection of the HIV env/rev retroviral vector was examined in rhesus macaques and baboons (Irwin et al. 1994). Animals were injected IM with 10^6–10^8 cfu/ml of HIV env/rev lyophilized vector preparations at multiple (4) injection sites and immune responses specific for HIV-1 env/rev were assessed prior to and after vector administration. Peripheral blood mononuclear cells (PBMCs) were obtained from vector-treated macaques and baboons and stimulated in vitro with HIV-1 env/rev expressing, vector transduced, *Herpes papio*-transformed, autologous B lymphoblastoid cell line (BLCL) cells. The vector-treated animals generated HIV-1 env/rev specific cytolytic activity as shown by lysis of transduced autologous target cells compared to nontransduced control targets. PBMCs obtained prior to vector injection and in vitro stimulated with transduced cells did not lyse target cells, indicating that in vivo priming was necessary for CTL induction in these animals. Therefore, the retroviral vector was capable of activating CTL in nonhuman primates similar to that observed in the murine model.

The effector cell population induced by HIV-env/rev vector administration in nonhuman primates was characterized for the CD4/CD8 phenotype of these T cells (IRWIN et al. 1994). T cell depletion studies indicated that both macaque and baboon effector cell populations were $CD8^+$, $CD4^-$ CTLs, as shown by abrogation of cytotoxic activity following depletion of $CD8^+$ T cells, whereas $CD4^+$ T cell depletion showed no reduction in the lysis of target cells. The CTLs from macaques and baboons were unable to lyse heterologous target cells suggesting that these cytotoxic responses were also self-MHC-restricted. Thus, the effector cells induced in nonhuman primates following retroviral vector immunization appear to be $CD8^+$, MHC-restricted CTLs.

Rhesus macaques were injected with varying amounts of HIV env/rev vector to determine dosing effects. Rhesus monkeys injected with 10^6 cfu of vector developed CTL activity, but the frequency of animals responding was low (Fig. 2). Injection of a higher 10^7 cfu dose of vector significantly increased the frequency of responding animals, although the magnitude of CTL activity in responding animals was similar. There appeared to be a dose-dependent threshold effect of vector dose on CTL activation in this model regarding frequency of responding animals. Monkeys given a booster injection of 10^7 cfu vector 16 weeks following initial immunization demonstrated CTL activation comparable to initial CTL activity. These results suggest repeated administration of the retroviral vector in monkeys is capable of reactivating CTL activity.

In studies with baboons, sera from 10^7 cfu vector-immunized animals have shown HIV-1 gp120-specific antibody (IgG) induction by ELISA. Rhesus macaques immunized with 10^8 cfu, but not 10^7 cfu vector doses also elicited moderate levels of anti-HIV-1 gp120 antibody (IgG). Vector transduction efficiency may vary within different animals or immune activation thresholds may require different antigen loads for immune responses to occur. Monkeys previously immunized with 10^7 cfu vector and boosted 16 weeks post-initial injection were found to elicit an augmented anti-HIVgp120 antibody (IgG) response. These results indicated that repeated administration of vector is feasible in these animals for inducing both humoral and cellular immunity. The dose level of the vector does appear to be important regarding the frequency of responding animals as well as the type of immune response activated (i.e., humoral or cellular).

4 Human Studies

Based on the previous studies of retroviral vector immunization in experimental animals, phase I clinical studies were initiated using retroviral vector treatment of HIV-infected individuals. These studies were designed to evaluate safety of the retroviral vector treatment and to establish biological activity with the eventual intent to develop an immunotherapeutic to augment cellular immune responses in HIV-seropositive subjects. The rationale for immune-based therapies is that the

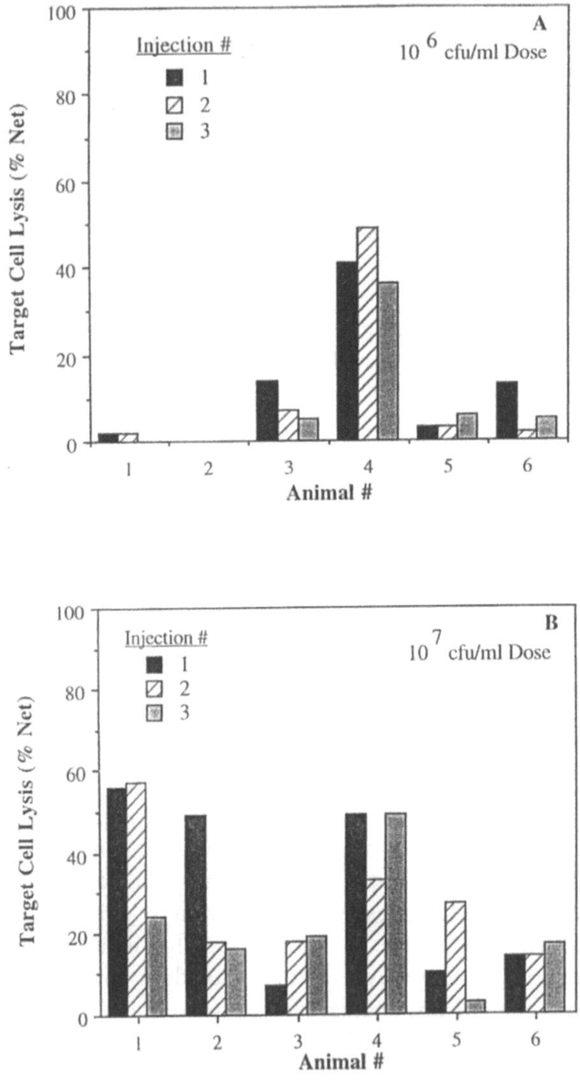

Fig. 2A, B. Cytotoxic T lymphocyte (CTL) activity in rhesus macaques following injection of the HIVenv/rev retroviral vector. Two groups of macaques were injected three times with **A** 10^6 or **B** 10^7 cfu/ml of vector at four intramuscular sites (0.5 ml/site). Peripheral blood mononuclear cells were obtained 2 weeks after each injection, stimulated in vitro with irradiated vector transduced autologous B lymphoblastoid cell lines (BLCL) expressing HIV-1 env/rev protein, and the resulting effector cells tested for CTL activity on autologous targets expressing HIV env/rev/neo^r (HIV env/rev). Effectors were also tested on control targets expressing β-galactosidase and neo^r (βgal). The results are expressed as the net difference in lysis between the HIV env/rev and βgal targets

natural immune response to virus infection is incomplete and can be augmented through optimal antigen presentation to achieve better control of virus infection and disease progression.

4.1 Human Cytotoxic T Lymphocyte Monitoring System

The CTL response may have an important role in controlling the HIV disease process. Gene transfer technology using retroviral vectors may not only have a potential treatment application, but could be useful for evaluating human CTL activity. To this end, studies were designed to develop a human CTL assay using Epstein-Barr virus (EBV)-transformed BLCL derived from HIV-infected subjects. The BLCL were transduced with the retroviral vector encoding different HIV proteins, i.e., env, rev, and gag. The intrinsic variability of the assay was established and longitudinal monitoring of CTL activity in HIV subjects was evaluated. Using the retroviral vector-transduced BLCL target cell system, we have demonstrated the ability to monitor CTL activity in multiple subjects over 1–2 years. CD8+, class I MHC-restricted CTL activity against HIV-1IIIB env/rev and gag targets was detected in most subjects with CD4 cell counts of 200–600 cells/mm^2. The use of continuous vector-transduced stimulator/target cell lines expressing appropriate HIV proteins was useful in standardizing the CTL assay. The results indicated that the retroviral vector-transduced target cells were capable of effectively presenting HIV determinants for CTL activation to PBMCs obtained from subjects likely infected with a wide variety of HIV isolates. Furthermore, the in vitro stimulated CTLs were capable of lysing HLA-matched, heterologous HIV-1 IIIB infected BLCL (Fig. 3). The CTL activity did not appear to be influenced by natural killer (NK) cell activity and the impact of EBV-specific CTL activity was reduced through the use of "cold" target inhibition. Therefore, the vector-transduced cells were capable of specifically activating CTLs that could subsequently recognize determinants expressed through the HIV infection process. An equivalent in vivo stimulation of CTLs could potentially impact the HIV disease process.

4.2 Retroviral Vector Clinical Studies

Initial phase I clinical studies were performed involving the IM administration of 1×10^7 ex vivo HIVenv/rev retroviral vector-transduced autologous fibroblasts (VTAF) in HIV-seropositive, asymptomatic patients. HIV-infected patients (i.e., three of four subjects) treated with VTAF have shown augmented HIV-1 IIIBenv/rev-specific CD8+, class I MHC-restricted CTL responses (Ziegner et al. 1995). No adverse clinical events were observed in these patients and the treatment with VTAF was well tolerated.

As an extension of these ex vivo studies, several phase I clinical studies involving the direct IM administration of the HIVenv/rev retroviral vector to HIV-seropositive, asymptomatic patients were initiated. These studies were designed to evaluate different dosing regimens in HIV-infected patients with varying CD4 cell counts.

The safety and biological activity of the HIVenv/rev retroviral vector immunization was evaluated in 21 HIV-seropositive individuals with CD4 > 400; the patients were randomized into four treatment groups: (1) placebo, (2) 10^6 cfu

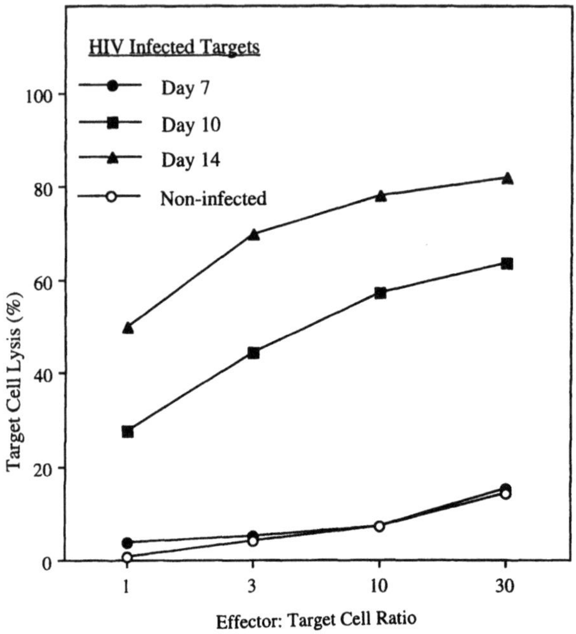

Fig. 3. Human cytotoxic T lymphocyte (CTL) lysis of HIV-1 IIIB infected target cells. Peripheral blood mononuclear cells obtained from HIV seropositive subjects were stimulated in vitro for 7–10 days with irradiated vector transduced B lymphoblastoid cell lines (BLCL) expressing HIV-1 IIIB env and rev proteins. The resulting effector cells were tested on HLA-matched targets infected with HIV-1 IIIB for 7, 10, and 14 days or on noninfected targets

vector administered at one IM site, (3) 10^6 cfu administered at four IM sites, and (4) 10^7 cfu administered at four IM sites. Injections were administered at 0, 30, and 60 days. CD4 cell counts, p24 antigen levels, HIV-1 env/rev-specific CTL activity, HIV RNA and DNA levels were monitored for up to 1 year following vector treatment. CTL activity was measured using fresh PBMCs from treated subjects that were stimulated in vitro with HIV env/rev expressing, irradiated autologous cells and assayed on autologous vector-transduced radiolabeled BLCL as target cells in a standard 5 h ^{51}Cr-release assay. Nonradiolabeled "cold" autologous BLCL (not expressing HIV env/rev) were added to the cytotoxicity assay to competitively inhibit EBV-directed CTL activity. The majority of subjects exhibited CTL reactivity against EBV-expressing target cells. There were no augmented HIV env/rev-specific CTL responses detected in the placebo and 10^6 cfu-1 injection site groups (Table 1). The 10^6 cfu-4 injection site group showed one CTL responder, whereas the 10^7 cfu-4 injection site group exhibited a higher frequency of CTL responders (4/5) based on differences between pre- and post-treatment CTL values. CTL activity changes from baseline of net HIV env/rev-specific CTL activity showed significant treatment group differences with the 10^7 cfu-4 injection site group ($p < 0.05$) at 120 days post-treatment. Median CD4 counts were stable as were HIV RNA and DNA levels in all study groups. The HIV env/rev retroviral

Table 1. Cytotoxic T lymphocyte (CTL) responses in HIV-seropositive subjects treated with different doses of HIV env/rev retroviral vector

	Placebo	10^6cfu	10^6cfu	10^7cfu
Injection sites	1	1	4	4
CTL responders	0/6	0/6	1/4	4/5
Total treated				21

Subjects were injected intramuscularly with vector at one or four injection sites. CTL activity was monitored over a 4 month study interval.

vector treatments were well tolerated and immunogenic at a monthly × 3 schedule at the 10^7 cfu-4 injection site dose level.

A subsequent phase I study involved three monthly IM injections of 10^7 cfu HIV env/rev vector adminisistered to 16 HIV-seropositive, asymptomatic subjects with CD4 counts 200–499 cells/mm^3 in one group and >500 CD4 counts in a second group. Six patients were on stable antiretroviral therapy. Clinical exams, standard lab tests, CD4 counts, CTL activity using fresh stimulated PBMCs tested on vector-transduced autologous BLCL targets, and PCR analysis of HIV viral RNA were monitored. Four months post-treatment, no significant clinical or laboratory toxicity related to study treatment was observed. The mean CD4 counts and slopes showed a nonsignificant increase following analysis of four pre- and four post-treatment values. HIV RNA levels were unchanged. Two of six patients with CD4 > 500 and three of ten patients with CD4 < 500 showed increases in CTL activity. The HIV env/rev retroviral vector treatment appears to be safe in HIV-seropositive, asymptomatic patients with CD4 > 200 and to be capable of increasing CTL activity in a subset of patients.

5 Discussion

The retroviral vector gene transfer system for HIV immunotherapy appears to be well tolerated in humans and to be feasible for stimulating immune responses, particularly CTL responses based on these experimental animal and human studies. A unique immunological aspect of gene transfer technology is the ability to provide antigen to the appropriate antigen processing pathways for efficient T cell activation. A number of gene transfer immunization strategies are currently being pursued using viral and nonviral gene delivery (Moss 1988; Robinson et al. 1993; Ulmer et al. 1993; Wang et al. 1993). These strategies appear to be all capable of activating both cellular and humoral immune responses. As greater understanding of gene transfer is achieved, the unique aspects of the different approaches could have application to number of disease situations.

CTL specific for HIV- 1 proteins have been demonstrated early in the course of HIV infection, prior to the appearance of antibody, and are believed to be involved in controlling the initial stages of the disease (Koup et al. 1994). Cells with the CTL phenotype have been shown to limit in vitro HIV replication (Walker et al. 1986) and are associated with the clearance of simian immunodeficiency virus (SIV) antigenemia in infected rhesus monkeys (Yasutomi et al. 1993). $CD8^+$ CTLs appear to be generated within a matter of days de novo following HIV or SIV infections. During the chronic course of HIV-1 infection, the temporal association of increases in viral load concomitant with decreases in CTL activity suggests that augmentation of HIV-specific CTL activity may have potential beneficial impact on disease progression (Connor et al. 1993; Koup et al. 1994; Riviere et al. 1989).

However, recent studies (Zinkernagel and Hengartner 1994; Koenig et al. 1995) have indicated the possibility that CTLs can contribute to pathogenesis when there is a highly focussed, single determinant-specific CTL response. In this case, cloned $CD8^+$ CTL specific for a single nef protein eptiope were infused ($\geq 10^{10}$ cells plus interleukin-2) into an HIV-seropositive patient and resulted in increases of viral load as well as decreases of CD4 cells at each dosing. These results suggested that highly specific CTLs may select for viral variants which could contribute to HIV pathogenesis. In addition, HIV-specific CTLs may be capable of producing cytokines that may pose concerns regarding stimulation rather than inhibition of virus replication. Although the pathogenesis of HIV-1 infection is not clearly understood, the possibility exists that boosting broadly specific cellular immune responses, particularly CTL activity, could be clinically useful in HIV-infected subjects, particularly in the early stages of the disease.

The changes in measured CTL activity in the retroviral vector-treated subjects could be due to a variety of factors given the degree of immune responsiveness and viral load in these patients. Certainly, the CTL assay in its current form is not quantitative and variation from assay to assay can be high. However, the comparison of multiple pre- and post-treatment CTL determinations has helped to provide greater confidence in the estimator of in vivo CTL activity. In these preliminary clinical studies, the doses and dose intervals employed may not be optimal to obtain either a high frequency of CTL responders or the duration of activity required to impact the disease process. In addition, the actual role of CTLs and the appropriate antigen(s) for recognition of patient viral isolates have not been well defined for a therapeutic to ensure a positive impact on HIV disease progression. Nonetheless, these studies have been critically important as the basis for the development of the retroviral vector-mediated gene transfer system for humans and may represent an initial step in defining the potential of immune stimulation for HIV disease treatment.

Acknowledgements. We thank Dr. Jeff Galpin, Shared Medical Research Foundation, Tarzana CA, and Dr. Richard Haubrich, UCSD Treatment Center, San Diego, CA, for conducting the clinical trials; Gloria Peters and the Clinical Response Evaluation Group for clinical testing, Carol Gay Anderson and the Human Cellular Immunology Group for human CTL assay development, Dr. Wolfgang Klump for clinical PCR testing, Dr. Mike Irwin, Dr. Jim McCormack and Dr. Mangala Hariharan for the murine and macaque studies, Dr. Sunil Chada, Dr. Elizabeth Song, and Duane Brumm for vector distribution studies, Dr. Kieron Kowal for vector production, and Dr. Minoru Hirama, Dr. Mas Nishida, and Dr. Steven Mento for support and useful discussions. This work was supported as a joint project between Green Cross Corporation, Osaka, Japan and Chiron Viagene.

References

Boshart M, Weber F, Jahn G, Dorsch-Hasler K, Fleckenstein B, Schaffner W (1985) A very strong enhancer is located upstream of an immediate early gene of human cytomegalovirus. Cell 41:521–530

Braciale TJ, Morrison LA, Sweetser MT, Sambrook J, Gething M-J, Braciale VL (1987) Antigen presentation pathways of class I and class II MHC-restricted T lymphocytes. Immunol Rev 98:95–114

Burns JC, Friedmann T, Driever W, Burrascano M, Yee JK (1993) Vesicular stomatitis virus G glycoprotein pseudotyped retroviral vectors: concentration to very high titer and efficient gene transfer into mammalian and nonmammalian cells. Proc Natl Acad Sci USA 90:8033–8037

Carmichael A, Jin X, Sissons P, Borysiewicz L (1993) Quantitative analysis of the human immunodeficiency virus type I (HIV-I)-specific cytotoxic T lymphocyte (CTL) response at different stages of HIV- I infection: differential CTL responses to HIV- I and Epstein Barr virus in late disease. J Exp Med 177:249–256

Chada S, Dejesus CE, Townsend K, Lee WTL, Laube L, Jolly DJ, Chang SMW, Warner JF (1993) Cross-reactive lysis of human targets infected with prototypic and clinical human immunodeficiency virus type I (HIV- 1) strains my murine anti-HIV-1 IIIB env-specific cytotoxic T lymphocytes. J Virol 67:3409–3417

Connor RI, Mohri H, Cao Y, Ho DD (1993) Increased viral burden and cytopathicity correlate temporally with CD4+ T-lymphocyte decline and clinical progression in human immunodeficiency virus type 1-infected individuals. J Virol 67:1772–1778

Germain RN, Margulies DH (1993) The biochemistry and cell biology of antigen processing and presentation. Annual Rev Immunol 11:403–450

Ho DD, Neumann AU, Perelson AS, Chen W, Leonard JM, Markowitz M (1995) Rapid turnover of plasma virions and CD4 lymphocytes in HIV- I infection. Nature 373:123–126

Hoffenbach A, Langlade-Demoyen P, Dadaglio G, Vilmer E, Michel F, Mayaud C, Autran B, Plata F (1989) Unusually high frequencies of HIV-specific cytotoxic T lymphocytes in humans. J Immunol 142:452–460

Irwin MJ, Laube LS, Lee V, Austin M, Chada S, Anderson C-G, Townsend K, Jolly DJ, Warner JF (1994) Direct injection of a recombinant retroviral vector induces human immunodeficiency virus-specific immune responses in mice and nonhuman primates. J Virol 68:5036–5044

Kast WM, Bronkhorst AM, De waal LP, Melief CJM (1986) Cooperation between cytotoxic and helper T lymphocytes in protection against lethal Sendai virus infection. J Exp Med 164:723–738

Klavinskis LS, Whitton JL, Oldstone MBA (1989) Molecularly engineered vaccine which expresses an immunodominant T-cell epitope induces cytotoxic T lymphocytes that confer protection from lethal virus infection. J Virol 63:4311–4316

Koenig S, Conley AJ, Brewah YA, Jones GW, Leath S, Boots LJ, Davey V, Pantaleo G, Dewarest JF, Carter C, Wannebo C, Yanneli JR, Rosenberg SA, Lane HC (1995) Transfer of HIV-I specific cytotoxic T lymphocytes to an AIDS patient leads to selection for mutant HIV variants and subsequent disease progression. Nature Med 1:330–336

Koup RA, Safrit JT, Cao Y, Andrews CA, Mcleod G, Borkowsky W, Farthing C, Ho DD (1994) Temporal association of cellular immune responses with the initial control of viremia in primary HIV-I syndrome. J Virol 68:4550–4559

Lander MR, Chattopadhyay SK (1984) A *Mus dunni* cell line that lacks sequences closely related to endogenous murine leukemia viruses and can be infected by ecotropic, amphotropic, xenotropic, and mink cell focus-forming viruses. J Virol 52:695–698

La Rosa GJ, Davide JP, Weinhold K, Waterbury JA, Profy AT, Lewis JA, Langlois AJ, Dreesman GR, Boswell RN, Shadduck P (1990) Conserved sequence and structural elements in the HIV- I principal neutralizing determinant. Science 249:932–935

Laube LS, Burrascano M, Dejesus CE, Howard BD, Johnson MA, Lee WTL, Lynn AE, Peters G, Ronlov GS, Townsend KS, Eason RL, Jolly DJ, Merchant B, Warner JF (1994) Cytotoxic T lymphocyte and antibody responses generated in rhesus monkeys immunized with retroviral vector-transduced fibroblasts expressing human immunodeficiency virus type-I IIIB *env/rev* proteins. Human Gene Therapy 5:853–862

Maddon PJ, Dalgleish AG, Mcdougal JS, Clapham PR, Weiss RA, Axel R (1986) The T4 gene encodes the AIDS virus receptor and is expressed in the immune system and the brain. Cell 47:333–348

Malim MH, Hauber J, Fenrick R, Cullen BR (1988) Immunodeficiency virus rev trans-activator modulates the expression of the viral regulatory genes. Nature 335:181–183

Moss (1988) Use of vaccinia virus vector for development of AIDS vaccines. AIDS S103–S105

Patek PQ, Collins JL, Cohn M (1982) Activity and dexamethasone sensitivity of natural cytotoxic cell subpopulations. Cell Immunol 72:113–121

Peebles PT (1975) An in vitro focus induction assay for xenotropic murine leukemia virus, feline leukemia virus C and the feline-primate viruses RD-I 14/CCC/M-7. Virology 67:288–291

Price J, Turner D, Cepko C (1987) Lineage analysis in the vertebrate nervous system by retrovirus-mediated gene transfer. Proc Natl Acad Sci USA 84:156–160

Printz M, Reynolds J, Mento SJ, Jolly D, Kowal K, Sajjadi N (1994) Recombinant retroviral vector interferes with the detection of amphotropic replication competent retrovirus in standard culture assays. Gene Ther 2:143–150

Reddehase MJ, Mutter W, Munch K, Buhring H-J, Koszinowski UH (1987) CD8-positive T lymphocytes specific for murine cytomegalovirus immediate-early antigens mediate protective immunity. J Virol 61:3102–3108

Reusser P, Riddel SR, Meyers JD, Greenberg PD (1991) Cytotoxic T-lymphocyte response to cytomegalovirus after human allogeneic bone marrow transplantation: pattern of recovery and correlation with cytomegalovirus infection and disease. Blood 78:13731380

Riviere Y, Tanneau-Salvadori F, Regnault A, Lopez O, Sansonetti P, Guy B, Kieny M-P, Fournel J-J, Montagnier L (1989) Human immunodeficiency virus-specific cytotoxic responses of seropositive individuals: distinct effector cells mediate killing of targets expressing *gag* and *env* proteins. J Virol 63:2270–2277

Robinson HL, Fynan EF, Webster RG (1993) Use of direct DNA inoculations to elicit protective immune responses. In: Vaccines 93 Cold Spring Harbor Laboratory, Cold Spring Harbor, NY, pp 311–315

Rouse RT, Norley S, Martin S (1988) Anti-viral cytotoxic T lymphocyte induction and vaccination. Rev Infect Dis 10:16–33

Sajjadi N, Kamantigue E, Edwards W, Howard T, Jolly D, Mento S, Chada S (1994) Recombinant retroviral vector delivered intramuscularly localizes to the site of injection in mice. Human Gene Therapy 5:693–699

Schrager LK, Young JM, Fowler MG, Mathieson BJ, Vermund SH (1994) Long-term survivors of HIV-I infection: definitions and research challenges. AIDS 8 (suppl 1):S95–S108

Takahashi H, Cohen J, Hosmalin A, Cease KB, Houghton R, Cornette JJ, DeLisa, C, Moss B, Germain R, Berzofsky JA (1988) An immunodominant epitope of the human immunodeficiency virus envelope glycoprotein gp 160 recognized by class I major histocompatibility complex molecule-restricted murine cytotoxic T lymphocytes. Proc Natl Acad Sci USA 85:3105–3109

Ulmer JB, Donnelly J, Parker SE, Rhodes GH, Felgner PL, Dwarki VL, Gromkowski SH, Deck R, Devitt CM, Friedman A, Hawe LA, Leander KR, Marinez D, Perry H, Shiver JW, Montgomery D, Liu MA (1993) Heterologous protection against influenza by injection of DNA encoding a viral protection. Science 259:1745–1749

Walker CM, Moody DJ, Stites DP, Levy JA (1986) CD8+ lymphocytes can control HIV infection in vitro by suppressing virus replication. Science 234:1563–1566

Wang B, Boyer J, Srikantan V, Coney L, Carrano R, Phan C, Merva M, Dang K, Agadjanyan M, Gilbert L, Ugen K, Williams VW, Weiner DB (1993) DNA inoculation induces neutralizing inunune responses against human immunodeficiency virus type I in mice and non-human primates. DNA Cell Biol 12:799–805

Warner JF, Anderson C-G, Laube L, Jolly DJ, Townsend K, Chada S, St Louis D (1991) Induction of HIV-specific CTL and antibody responses in mice using retroviral vector-transduced cells. AIDS Res Hum Retroviruses 7:645–655

Wei X, Ghosh SK, Taylor ME, Johnson VA, Emini EA, Deutsch P, Lifson JD, Bonhoeffer S, Nowak MA, Hahn BH, Saag MS, Shaw GM (1995) Viral dynamics in human immunodeficiency virus type I infection. Nature 373:117–122

Yasutomi Y, Reimann KA, Lord CI, Miller MD, Letvin NL (1993) Simian immunodeficiency virus-specific CD8+ lymphocyte response in acutely infected rhesus monkeys. J Virol 67:1701–1711

Ziegner UHM, Peters G, Jolly DJ, Mento SJ, Galpin J, Prussak CE, Barber JR, Hartnett DE, Bohart C, Klump W, Sajjadi N, Merchant B, Warner JF (1995) Cytotoxic T-lymphocyte induction in asymptomatic HIV-1-infected patients immunized with Retrovector-transduced autologous fibroblasts expressing HIV-1IIIB *env/rev* proteins. AIDS 9:43–50

Zinkernagel RM, Hengartner H (1994) T-cell-mediated immunopathology versus direct cytolysis by virus: implication for HIV and AIDS. Immunol Today 15:262–268

Developing DNA Vaccines Against Immunodeficiency Viruses

S. Lu

1 Introduction

More than a decade after the identification of human immunodeficiency virus-1 (HIV-1) as the causative pathogen of AIDS, the prospects for an HIV-1 vaccine in the foreseeable future are still low. Attempts using conventional vaccine approaches have not met with much success (for recent reviews see Haynes 1993; Bolognesi 1993, 1995). The live attenuated vaccine approach appeared promising when its effectiveness was demonstrated in a simian immunodeficiency virus (SIV) model in macaques (Daniel et al. 1994, 1992). However, additional studies in neonatal macaques raised safety concerns regarding this approach (Baba et al. 1995). The search for novel vaccine strategies that target the unique biological features of HIV-1 remains a challenging yet important task.

DNA-based vaccines, with their ability to express antigens in vivo, have the great potential to become an ideal vaccination approach against HIV-1. There are at least three strong arguments in favor of such an HIV-1 vaccine. First, effective neutralizing antibodies against envelope proteins of HIV-1 are often conformation-dependent (Haigwood et al. 1992; Javaherian et al. 1994). A DNA-based vaccine can almost mimic the live attenuated vaccine approach to produce native viral proteins or even noninfectious viral particles in vivo. At the same time, the pathogenic potential from the original viral genome can be selectively eliminated to

Department of Medicine, University of Massachusetts Medical Center, Worcester, MA 01655, USA make this approach much safer than a live-attenuated vaccine. Second, virus-specific cytotoxic T lymphocytes (CTLs) play an important role in containing the early spread of immunodeficiency virus infections (Koup et al. 1994; Yasutomi et al. 1993). But many of the HIV-1 vaccine strategies including the use of recombinant HIV-1 envelope glycoproteins as immunogens do not elicit virus-specific CTL (Letvin 1993). Since class I MHC-restricted presentation of viral antigens occurs most efficiently in the course of intracellular synthesis of viral proteins, DNA-based vaccines can easily access this intracellular pathway to prime the CTL arm of the immune system. Finally, HIV-1 has made high variability its trademark, i.e., viral isolates with new mutations or even new strains are generated during the course of infection, both in one individual and in a population. Even though we do not know at this time what strains should be used as our critical targets in controlling the spread of infection, it is conceivable that future HIV-1 vaccines will be formulated on a polyvalent basis. Because DNA-based vaccines for different HIV-1 strains can be easily produced and inoculated as a mixture, the efficacy of these vaccines can be quickly tested and the cost of doing these tests will be lower than that using recombinant or live-attenuated vaccine approaches. This becomes even more important when the HIV-1 vaccines are tested in large-scale clinical trials, especially in third world countries.

Since its inception, DNA vaccination has been shown to generate immune responses against a number of viruses including influenza A viruses (Fynan et al. 1993; Ulmer et al. 1993), bovine herpesvirus (Cox et al. 1993), hepatitis B virus (Davis et al. 1993), rabies virus (Xiang et al. 1994) and lymphocytic choriomeningitis virus (Zarozinski et al. 1995; Martins et al. 1995). Several groups have independently reported successful results generating HIV-1-specific antibody and CTL responses in mice and nonhuman primates (Wang et al. 1993a, b; Fuller and Haynes 1994; Lu et al. 1995, 1996). These studies confirmed DNA vaccine as a useful, novel approach to raise humoral and cell-mediated immune responses against HIV-1. The efficacy of DNA vaccines to protect against immunodeficiency viruses was also studied in non-human primate models (Wang et al. 1995b; Lu et al. 1996). This review attempts to give a brief summary of our recently concluded SIV DNA vaccine trial against a virulent, uncloned SIVmac251 challenge in macaques (Lu et al. 1996; Yasutomi et al. 1996). It is hoped that discussion of the following findings from this trial will help us to focus our effort for the final development of a DNA vaccine against HIV-1.

2 Can Current DNA Vaccines Confer Complete Protection Against Immunodeficiency Viruses?

After the ability of DNA vaccines to raise immune responses against HIV-1 antigens was established initially in mice, the central question regarding the development of a DNA-based HIV-1 vaccine is whether the current DNA vaccines can provide protection against the infection of a viral challenge in an animal model. It was first demonstrated that mice immunized with DNA vaccine expressing HIV-1 gp160 could reject the growth of otherwise lethal tumor cells which were transfected with HIV-1 Env-expressing vector (WANG et al. 1995a). But in a primate study using the similar vaccine design, all of four immunized cynomolgus macaques became clearly infected after the challenge of SHIV, a chimeric of HIV-1 and SIV. However, three of them did exhibite lower levels of viral load than the control macaques (WANG et al. 1995b).

We started our macaque trial to test the strength of SIV DNA vaccines against a virulent, uncloned SIVmac251 challenge (Fig. 1). Our DNA vaccines generated SIVmac-specific CTL and antibody responses but failed to achieve full protection against the challenge (LU et al. 1996). In this study, rhesus monkeys (*Macaca mulatta*) were inoculated with an experimental vaccine consisting of five DNA plasmids expressing different combinations and forms of SIVmac proteins. One vaccine plasmid encoded nonreplicating SIVmac239 virus particles. The other four plasmids encoded secreted forms of the envelope glycoproteins (gp110 and gp130 of SIVmac239, gp130 of SIVmac251 and gp130 of SIVmac316). SIVmac239 and SIVmac251 are T cell tropic relatives and SIVmac316 is the monocyte/macrophage tropic relative of the uncloned challenge. After a total of six DNA immunizations at weeks 1, 3, 11, 13, 21, and 23, each of the animals was challenged intravenously at week 25 with 10 monkey infectious doses of uncloned SIVmac251 (Fig. 1).

The DNA immunizations did not prevent early infection or protect against later CD4+ cell loss (LU et al. 1996). Long-term chronic levels of infection were similar in the vaccinated and control animals. However, viral loads were reduced to the chronic level over a shorter period of time in the vaccinated macaques (6 weeks) than in the control macaques (12 weeks) (Fig. 2). Vaccinated animals also had a lower level of SIV p27 antigenemia during the first 6 weeks post-challenge (Fig. 3). The high p27 level in control macaques dropped to become similar to the level in the vaccinated macaques by week 8 (Fig. 3). Although the detection of plasma p27

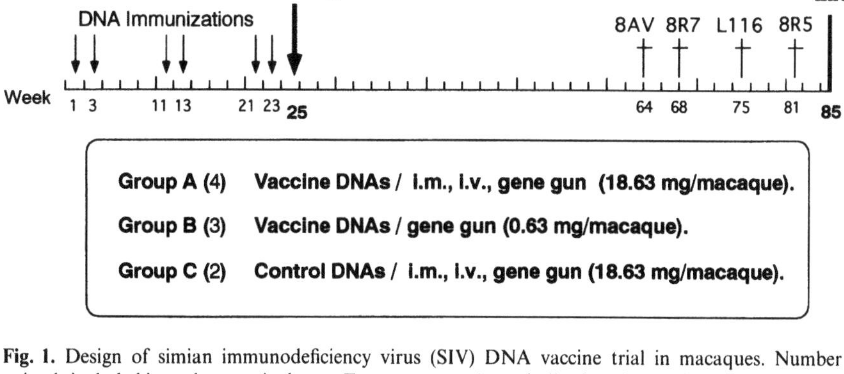

Fig. 1. Design of simian immunodeficiency virus (SIV) DNA vaccine trial in macaques. Number of animals included in each group is shown. Four macaques (*cross* indicating time of death) developed AIDS in the first year after receiving a SIVmac251 challenge

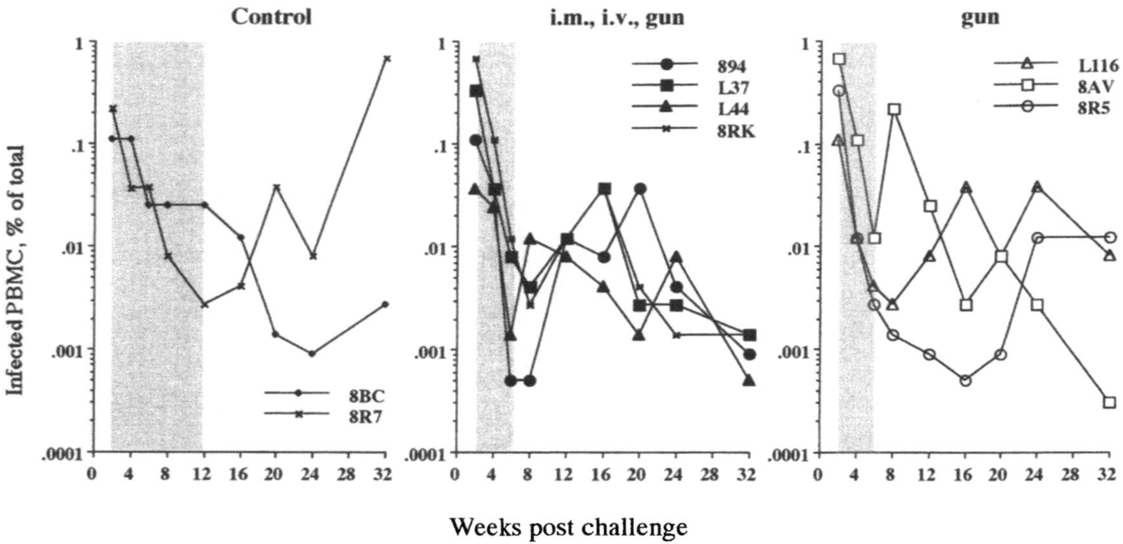

Fig. 2. Post-challenge, infected macaque peripheral blood mononuclear cells (PBMCs) as a percentage of total PBMCs. Cocultivations of PBMC with CEMx174 indicator cells were done by limiting dilution to determine the percentage of infected PBMC at various times post-challenge (Lu et al. 1996). *Shaded areas* represent the time required for the percentage of infected PBMCs to drop to the long-term post-infection level

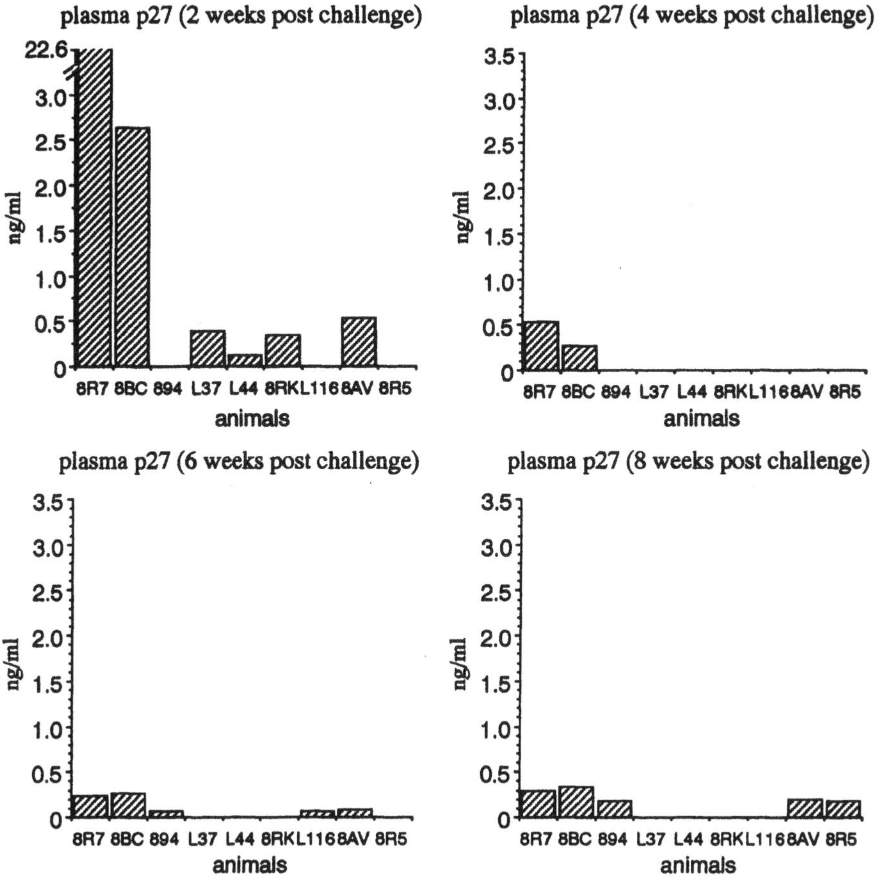

Fig. 3. Post-challenge antigenemia as shown by the levels of plasma p27. p27 antigen was assayed using the Coulter antigen capture ELISA (Coulter Corporation, Hialeah, FL). Note: the p27 level of macaque 8R7 was 22.6 ng at 2 weeks post-challenge. Assignment of animals: 8R7 and 8BC, control group; 894, L37, L44 and 8RK, 3-route vaccine group; L116, 8AV and 8R5, gene gun group

in this assay can be affected by the presence of anti-Gag antibody (Burns and Desrosiers 1992), our data may well be the result of a quick clearance of virus in immunized animals. It has been reported that antibody produced in response to vaccination could facilitate complement receptor 1 (CR-1)-mediated clearance of HIV-1 immune complexes through the mononuclear phagocytic system (Montefiori et al. 1994). The level of viremia was tested by incubating CEMx174 cells with macaque plasma at different dilutions; but even as early as 2 weeks post challenge, the level of viremia, as a result of natural clearance, was too low for us to detect the differences between vaccinated and control macaques (data not shown). Thus, the current formulations of DNA vaccines provided some attenuation of the acute phase of infection but did not prevent the loss of CD4+ cells (Lu et al. 1996).

Three of seven immunized and one of two control macaques developed AIDS-like symptoms within the first year of viral challenge (Fig. 1). The remaining animals were terminated at the end of trial (60 weeks post-challenge). It is interesting to note that three animals in the gene gun-only group developed opportunistic infections prior to the termination of the trial, whereas the four macaques receiving DNA vaccines by three routes (intramuscularly, intravenously, and by a gene gun) had been relatively healthy despite the drop of CD4+ cells. This difference in survival did not correlate with differences in antibody and CTL responses, differences in levels of post-challenge infection, or differences in the rate of CD4+ cell decline (Lu et al. 1996). Therefore, the difference in survival could have been due purely to chance. But in light of the recent report that HIV-1 virus trapped onto follicular dendritic cells (FDCs) in the germinal centers of lymphoid tissue remains a major infectious source (Heath et al. 1995), it is not clear whether animals vaccinated by three routes may have a lower virus load in FDCs. This may be the reason for the slow disease progress in these animals. Alternatively, the three-route approach may reduce the viral immunopathogenesis by mechanisms unknown to us at this time.

Failure to provide a complete protection against a virulent SIV challenge in macaques is quite a contrast with the successful protection against the infections of influenza and rabies in mice (Fynan et al. 1993; Ulmer et al. 1993; Xiang et al. 1994). We do not know whether the infection of immunodeficiency virus is more difficult to protect against or whether the DNA vaccines become less effective in primates. It is therefore important to confirm the effectiveness of DNA vaccines in primates by starting more protection trials in nonhuman primates against infections caused by pathogens other than HIV-1 and SIV. The increasing number of human DNA vaccine trials will certainly provide valuable information to guide the future development of human DNA vaccines.

3 Animal Models Play an Important Role in Determining the Efficacy of HIV-1 DNA Vaccines

Historically, morbidity and mortality of a given primate model for immunodeficiency virus vaccine studies depends on the virulence of the infectious agent (Hirsch and Johnson 1994; Bolognesi 1995). A number of primate models have been established for immunodeficiency virus vaccine trials, including HIV-1 (HIV-1, HIV-2), SIV (SIVmac, SIVmne, SIVsm), or hybrids of HIV-1 and SIV (SHIVs) (Li et al. 1992; Sakuragi et al. 1992). For the avirulent infections, such as HIV-1 in chimpanzees, HIV-2 in macaques, or SHIV in macaques, the infected primates usually control the infection and do not develop disease. Therefore, protective immunization has been relatively easy to achieve in these models with such vaccination approaches as recombinant vectors, envelope subunits, and even envelope peptides (Shafferman et al. 1991; Putkonen et al. 1991, 1994; Giard et al. 1991). In more virulent models, such as SIVmne infection in pig-tailed macaques, animals may develop disease but only after a latency period. For these animals, protection can be achieved only by the combination of vaccines using recombinant vaccinia vectors plus subunit boosts (Hu et al. 1993). Finally, in models that use highly virulent viruses, such as molecularly cloned SIVmac239 or uncloned SIVmac 251 infections in rhesus macaques, a high percentage of macaques develop AIDS during the first year of infection. In these highly pathogenic models, a prolonged infection with live attenuated virus has been required to generate protective immunity (Daniel et al. 1992; Almond et al. 1995). The fact that the success of a particular vaccine was dependent on the virulence level of the challenge model raised an important question: Which model should be used to test the efficacy of a DNA-based vaccine approach?

In our first SIV vaccine trial, the highly virulent uncloned SIVmac251/rhesus macaque model was chosen so that we could determine if a DNA vaccine could provide a non-infectious mimic of a live attenuated vaccine (Daniel et al. 1992). In this study, the DNA vaccines were subjected to a stringent test because the uncloned SIVmac 251 challenge virus is a "difficult-to-neutralize" stock. About half of the macaques challenged with this stock had developed AIDS-like diseases during the first year of infection. Our results demonstrated that the DNA vaccination approach used in this trial did not generate the protective responses achieved by the live attenuated vaccine approach. There are several important differences between these two approaches. First, DNA vaccinations in our trial were conducted over a 6 month period whereas it is now known that individuals with live attenuated infections require more than 1 year to develop protective responses against the same challenge stock used in our trial (Wyand et al. 1996). It is possible that DNA vaccination also requires a longer period to achieve the full maturation of a protective response. Second, unlike the live attenuated infection, the SIV DNA vaccines in this trial failed to induce persisting antibody responses (see below). However, it is not clear whether this failure affected the outcome of the trial because pre-challenge titers of neutralizing antibody did not correlate with the ability to protect against the difficult-to-neutralize, uncloned SIVmac251 challenge stock (Lu et al. 1996).

Because there is no ideal animal model in which to test the immunodeficiency virus vaccine and because it is still unknown what correlates are required to achieve a full protection in humans, it would be valuable to ask if DNA vaccines could actually protect populations undergoing rapid spread of AIDS by curtailing the acute phase of infections as shown in our first SIV trial in macaques. AIDS vaccines that attenuate the acute phase of infection may be more achievable than vaccines designed to prevent AIDS. At the same time, we need to search for better animal models which are more clinically relevent. One area of particular interest would be the ability of DNA vaccines to protect against HIV-1 infections at the mucosal surfaces. Studies from rotavirus DNA vaccines have shown that DNA vaccines are very effective in inducing protective immune responses against mucosal viral infections (Herrmann et al. 1996).

4 Broad But Less Persistent Antibody Responses Against Immunodeficiency Viruses

Significant ELISA and neutralizing antibodies were induced against SIVmac in all of the vaccinated but not the control macaques (Lu et al. 1996). The first cluster of DNA inoculations (at weeks 1 and 3) raised anti-Env IgG responses in two out of seven vaccinated macaques (29%) as shown by ELISA. The second cluster of inoculations (at weeks 11 and 13) was followed by a 100% seroconversion (all seven vaccinated macaques) as shown by both ELISA and neutralization assays. The neutralizing activity ranged from 1:216 to 1:768. However, anti-Env antibody responses were transient, with the titers of both ELISA and neutralizing activity falling between the second cluster (weeks 11 and 13) and the third cluster (weeks 21 and 23) of DNA inoculations. The third cluster of inoculations boosted the ELISA titers to levels similar to those achieved after the second cluster of inoculations, but failed to boost the neutralizing antibody titers which continued to fall (Lu et al. 1996).

The transience of anti-Env antibody responses in macaques is consistent with our earlier HIV-1 DNA vaccine study in Balb/C mice (Lu et al. 1995). In that study, DNA inoculations raised transient levels of antibody responses to Env but there were persistent levels of antibody to p24 virion capsid protein (CA). Because both Env and CA antigens were encoded in the same DNA vaccine construct, the data suggested that the persistence of antibody responses primarily depend on the immunogenicity of individual antigens. But the transient antibody response is not unique to Env of immunodeficiency viruses. It has also been seen in mice immunized with DNA vaccines expressing the circumsporozoite protein of malaria (Mor et al. 1995). These transient, humoral responses are in contrast to the persistent antibody responses seen in the study of mice immunized with DNA vaccines expressing the influenza hemagglutinin antigen (Robinson et al. 1995). Finally, the transient antibody responses are independent of the route of DNA inoculation (Lu et al., 1995, 1996). In our HIV-1 DNA vaccine study in mice, we started with intramuscular and intravenous inoculations. Once the antibody responses reached the peak level and started falling, additional inoculations including gene-gun delivery failed to boost the antibody responses. In our macaque trial, macaques in the three-route (im, iv, gene-gun) group and the gene-gun-only group basically had the same temporal pattern of antibody responses, both falling after the second cluster of inoculations. Thus, the persistency of antibody appeared to be a function of the expressed antigen. The poor immunogenicity of HIV-1 env may be the result of its heavily glycosylated, oligomeric structures.

Despite the failure of the third cluster of DNA inoculations to boost neutralizing activity, the challenge infection resulted in high titers of neutralizing antibody for SIVmac251. At 1 month post challenge, the antibody titers in the vaccinated groups were similar to or higher than those in the control group. Thus, DNA vaccination clearly did not induce an immunological anergy status in these animals. On the contrary, DNA vaccines successfully primed a broad neutralizing response because the vaccinated animals had much higher titers of neutralizing activity for SIVsmB670 than the control animals 1 month post-viral challenge. It is not clear whether this early rise of a broad antibody response may have contributed to the quicker clearance of virus in vaccinated macaques soon after the challenge (Fig. 2). It is also unknown to us why macaques vaccinated with SIVmac239 antigens had a higher neutralizing titer against a distant SIV strain (SIVsmB670) than the closely related SIVmac251, but it is possible that SIVsmB670 is an easier viral strain to neutralize than uncloned SIVmac251.

Optimizing the dosing schedule of DNA vaccination and improving the level of antigen expression are important in boosting antibody responses to higher titers. Our macaque trial used a series of three clusters of DNA inoculations with intervening 8 week rest periods (each cluster consisted of two inoculations separated by 2 weeks). However, 4–5 month rest periods have been found to enhance the effect of protein boost in other immunodeficiency virus vaccine trials in primates (ANDERSON et al. 1989; GRAHAM et al. 1994). The clustered inoculations do not appear to have the marked effect on efficacy that was observed for clustered inoculations of low levels of protein antigens (DRESSER 1986) and, therefore, new immunization schedules with a longer resting period should be used in our future DNA vaccine trials. This is one of the examples that DNA-based immunization appears to obey some, but not all, of the principles that dictate the classical immunization schedule using protein antigens. At the same time, the ease of DNA vaccination allows us to combine DNA immunization with other immunological methods to further boost the antibody responses against immunodeficiency viruses. Co-inoculating cytokine-expressing plasmids or including adjuvants as part of the immunization schedule probably will help to achieve this goal.

In addition to raising ELISA and neutralizing activity, the DNA inoculations in our SIV trial also elicited complement-dependent enhancing antibody in pre-challenge sera of vaccine groups (LU et al. 1996). This enhancing activity could have contributed to the failure of the vaccines to prevent infection. In the post-challenge sera, both vaccine and control groups had similar levels of complement-dependent enhancing antibody. Therefore, DNA vaccine did not prime titers of enhancing antibody higher than those raised by natural infections. Enhancing assays on a series of stocks from molecularly cloned SIVmac239 suggest that different stocks of a single virus will not show the differences in susceptibility to enhancing antibody that are seen for susceptibility to neutralization activity (Montefiore, personal communication). This may reflect a higher level of conservation of epitopes that are targets for enhancement than of neutralization epitopes.

5 DNA Vaccines are Effective in Generating SIVmac Env- and Gag-Specific Cytotoxic T Lymphocyte Responses

SIVmac Env-specific CTL responses were detected in peripheral blood lymphocytes (PBLs) of all seven DNA-vaccinated macaques, and SIVmac Gag-specific CTL activity was demonstrated in PBLs of two of the seven vaccinated macaques (Lu et al. 1996; Yasutomi et al. 1996). The lower incidence of Gag-specific CTL than of Env-specific CTL responses may reflect the fact that all five vaccine DNAs expressed Env but only one expressed Gag, suggesting a suboptimal dose to induce a Gag-specific CTL response. This hypothesis is further supported by the finding that both macaques with Gag-specific CTL were in the three-route group (each macaque received a total of 18.6 mg DNA) whereas each macaque in the gene-gun-only group received 0.6 mg DNA (Fig. 1).

Both Env- and Gag-specific CTL responses were detected after the second cluster of inoculation. The levels of CTL activity were not further increased by the third cluster. In contrast to the antibody responses that fell with time, CTL responses were largely persistent (Lu et al. 1996; Yasutomi et al. 1996). Phenotype and MHC restriction analysis of the DNA vaccine-elicited effector cells indicated that they were CD8+, MHC class I-restricted CTLs (Yasutomi et al. 1996).

Success in raising Gag-specific CTLs in macaques, as well as in raising anti-Gag antibody in an HIV-1 DNA vaccine trial in mice (Lu et al. 1995) through in vivo production of noninfectious retroviral particles, may have an important impact on the future design of novel vaccines. In both trials, we have built "dpol" plasmids by selectively deleting the pol gene from HIV-1 or SIVmac genome. Both LTRs were also removed. Transfection of these dpol constructs into COS cells has resulted in the formation of noninfectious viral particles (Lu et al. 1995, 1996). The HIV-1.dpol and HIV-1.env DNAs showed no differences in their abilities to induce anti-Env antibodies, especially when examined by ELISA (Lu et al. 1995). But further studies are needed to clarify whether dpol-like constructs can induce higher neutralizing antibodies against conformational epitopes. Antigen expression through a dpol construct should maximize the chance to raise CTL responses in a population with broad MHC background and to raise antibody responses against additional viral proteins other than Env.

The data from our SIV DNA vaccine trial as summarized in this brief review clearly demonstrate that DNA vaccines were able to induce both antibody and CTL responses to immunodeficiency viruses and to suppress early SIV infection in macaques. Additional work is needed to further understand the pathogenesis of HIV-1 infection and new HIV-1 DNA vaccines should be designed to target at the critical conformation of HIV-1 virus which can raise effective neutralizing antibodies against the primary isolates. So far our HIV-1 or SIV DNA vaccines were produced by mainly following the similar designs of recombinant HIV-1 vaccines. However, these recombinant vaccines were not effective in inducing protective immune responses. Future development of effective HIV-1 DNA vaccines will depend on how much we can improve the immunogenicity of DNA vaccines to raise more persistent and stronger neutralizing antibody responses against different HIV-1 patient isolates.

Acknowledgements. The author is grateful to Dr. H Robinson for her leading role in our DNA vaccine studies discussed in this review and would like to thank her for discussion of this manuscript. The author thanks Dr. D. Montefiori for helpful discussion and Drs. J. Herrmann and T. Frado for critical comments on the manuscript.

References

Anderson KP, Lucas C, Hanson CV, Londe HF, Izu A, Gregory T, Ammann A, Berman PW, Eichberg JW (1989) Effect of dose and immunization schedule on immune response of baboons to recombinant glycoprotein 120 of HIV-1. J Inf Dis 160:960–969

Almond N, Kent K, Cranage M, Rud E, Clarke B, Stott EJ (1995) Protection by attenuated simian immunodeficiency virus in macaques against challenge with virus-infected cells. Lancet 345:1342–1344

Baba TW, Jeong YS, Penninck D, Bronson R, Greene MF, Ruprech RM (1995) Pathogenicity of live, attenuated SIV after mucosal infection of neonatal macaques. Science 267:1820–1825

Bolognesi DP (1993) Human immunodeficiency virus vaccines. Adv Virus Res 42:103–148

Bolognesi DP (1995) Ch. 20: The dilemma of developing and testing AIDS vaccines. In: Copper GM, Temin RG, and Sugden B (eds) The DNA provirus: Howard Temin's scientific legacy. American Society for Microbiology, Washington, DC, pp 301–312

Burns DPW, Desrosiers RC (1992) A caution on the use of SIV/HIV gag antigen detection system in neutralization assays. AIDS Res Hum Retroviruses 8:1189–1192

Cox GJM, Zamb TJ, Babiuk LA (1993) Bovine herpesvirus 1: immune responses in mice and cattle injected with plasmid DNA. J Virol 67:5664–5667

Davis HL, Michel M-L, Whalen RG (1993) DNA-based immunization induces continuous secretion of hepatitis B surface antigen and high levels of circulating antibody. Human Mol Gen 2:1847–1851

Daniel MD, Kirchhoff F, Czajak SC, Simon MA, Sehgal PK, Desrosiers RC (1992) Protective effects of a live attenuated SIV vaccine with a deletion in the nef gene. Science 258:1938–1941

Daniel MD, Mazzara GP, Simon MA, Sehgal PK, Kodama T, Panicali DL, Desrosiers RC (1994) High-titer immune responses elicited by recombinant vaccinia virus priming and particle boosting are ineffective in preventing virulent SIV infection. AIDS Res Hum Retroviruses 10:839–851

Dresser DW (1986) Immunization of experimental animals. In Weir DM (ed.): Immunochemistry, Handbook of Experimental Immunology, Blackwell Scientific Publications, Oxford, 1:8.1–8.21.

Fuller DH, Haynes JR (1994) A qualitative progression in HIV type 1 glycoprotein 120-specific cytotoxic cellular and humoral immune responses in mice receiving a DNA-based glycoprotein 120 vaccine. AIDS Res Hum Retroviruses 10:1433–1441

Fynan EF, Webster RG, Fuller DH, Haynes JR, Santoro JC, Robinson HL (1993) DNA vaccines: protective immunizations by parenteral, mucosal and gene-gun inoculations Proc Natl Acad Sci USA 90:11478–11482

Giard M, Kieny M-P, Pinter A, Barre-Sinoussi F, Nara P, Kolbe H, Kusumi K, Chaput A, Reinhart T, Muchmore E, Ronco J, Kaczorek M, Gomard E, Gluckman J-C, Fultz PN (1991) Immunization of chimpanzees confers protection against challenge with human immunodeficiency virus. Proc Natl Acad Sci USA 88:542–546

Graham BS, Gorse GJ, Schwartz DH, Keefer MC, McElrath MJ, Matthews TJ, Wright PF, Belshe RB, Clements ML, Dolin R, Corey L, Bolognesi DP, Sin DM, Esterlitz JR, Hu S-L, Smith GE and NIAID AIDS vaccine clinical trials network. (1994) Determinants of antibody response after recombinant gp160 boosting in vaccinia-naive volunteers primed with gp160-recombinant vaccinia virus. J Infect Dis 170:782–786

Haigwood NL, Nara PL, Brooks E, Van Nest GA, Ott G, Higgins KW, Dunlop N, Scandella CJ, Eichberg JW, Steimer KS (1992) Native but not denatured recombinant human immunodeficiency virus type 1 gp120 generates broad-spectrum neutralizing antibodies in baboons. J Virol 66:172–182

Haynes BF (1993) Scientific and social issues of human immunodeficiency virus vaccine development. Science 260:1279–1286

Heath SL, Tew JG, Tew JG Szakal AK, Burton GF (1995) Follicular dendritic cells and human immunodeficiency virus infectivity. Nature 377:740–744

Herrmann JE, Chen SC, Fynan EF, Santoro JC, Greenberg HB, Robinson HL (1996) DNA vaccines against rotavirus infections. Arch Virol (in press)

Hirsh VM, Johnson PR (1994) Pathogenic diversity of siman immunodeficiency viruses. Virus Research 32:183–203

Hu S-L, Stallard V, Abrams K, Barber GN, Kuller L, Langlois AJ, Morton WR, Benveniste RE (1993) Protection of vaccinia-primed macaques against SIVmac infection by combination immunization with recombinant vaccinia virus and SIVmne gp160. J Med Primatol 22:92–99

Javaherian K, Langlois AJ, Montefiori DC, Kent KA, Ryan KA, Wyman PD, Stott J, Bolognesi DP, Murphey-Corb M, LaRosa GJ (1994) Studies of the conformation-dependent neutralizing epitopes of simian immunodeficiency virus envelope protein. J Virol 68:2624–2631

Koup RA, Safrit JT, Cao Y, Andrews CA, Wu Y, McLeod G, Borkowsky W, Farthing C, Ho DD (1994) Temporal association of cellular immune response with the initial control of viremia in primary HIV-1 syndrome. J Virol 68:4650–4655

Li J, Haseltine W, Letvin NL, Sodroski J (1992) Infection of cynomolgus monkeys with a chimeric HIV-1/SIVmac virus that express the HIV-1 envelope glycoprotein. J AIDS 5:639–664

Letvin NL (1993) Vaccines against the human immunodeficiency viruses: progress and prospects. N Engl J Med 329:1400–1405

Lu S, Santoro JC, Fuller DH, Haynes JR, Robinson HL (1995) Use of DNAs expressing HIV-1 Env and noninfectious HIV-1 particles to raise antibody responses in mice. Virology 209:147–154

Lu S, Arthos J, Montefiori DC, Yasutomi Y, Manson K, Mustafa F, Johnson E, Santoro JC, Wissink J, Mullins JI, Haynes JR, Letvin NL, Wyand M, Robinson HL (1996) Simian immunodeficiency virus DNA vaccine trial in macaques. J Virol 70:3978–3991

Martins LP, Lau LI, Asano MS Ahmed R (1995) DNA vaccination against persistent viral infection. J Virol 69:2574–2582

Montefiori DC, Graham BS, Zhou JY, Zhou JT, Ahearn JM, and the National Institutes of Health AIDS Vaccine Clinical Trials Network (1994) Binding of human immunodeficiency virus 1 to the C3b/C4b receptor CR1 (CD35) and red blood cells in the presence of envelope-specific antibodies and complement. J Infect Dis 170:429–32

Mor G, Klinman DM, Shapiro S, Hagiwara E, Sedegah M, Norman JA, Hoffman SL, Steinberg AD (1995) Complexity of the cytokine and antibody response elicited by immunizing mice with Plasmodium yoelii circumsporozoite protein plasmid DNA. J Immunol 155:2039–2046

Putkonen P, Thorstensson R, Walther L, Albert J, Akerblom L, Granquist O, Wadell G, Norrby E, Biberfeld G (1991) Vaccine protection against HIV-2 infection in cynomolgus monkeys. AIDS Res Hum Retroviruses 7:271–277

Putkonen P, Nilsson C, Walther L, Ghavamzadeh L, Hild K, Broliden K, Biberfeld G, Thorstensson R (1994) Efficacy of inactivated whole HIV-2 vaccines with various adjuvants in cynomolgus monkeys. J Med Primatol 23:89–94

Robinson HL, Feltquate DM, Morin MJ, Haynes JR, Webster RG (1995) DNA vaccines: a new approach to immunization. In: Chanock RM, Brown F, Ginsberg HS and Norrby E (eds) Vaccines 95. Cold Spring Harbor Laboratory, Cold Spring Harbor, pp 69–75

Sakuragi S, Shibata R, Mukai R, Komatsu T, Fukasawa M, Sakai H, Saduragi J-I, Kawamura M, Ibuki K, Hayami M, Adachi A (1992) Infection of macaque monkeys with a chimeric human and simian immunodeficiency virus. J Gen Virol 73:2983–2987

Shafferman A, Jahrling PB, Benveniste RE, Lewis MG, Phipps TJ, Eden-McCutchan F, Sadoff J, Eddy GA, Burke DS (1991) Protection of macaques with a simian immunodeficiency virus envelope peptide vaccine based on conserved human immunodeficiency virus type 1 sequences. Proc Natl Acad Sci USA 88:7126–7130

Ulmer JB, Donnelly JJ, Parker SE, Rhodes GH, Felgner PL, Dwarki VJ, Gromkowski SH, Deck RR, DeWitt CM, Friedman A, Hawe LA, Leander KR, Martinez D, Perry HC, Shiver JW, Montgonery DL, Liu MA (1993) Heterologous protection against influenza by injection of DNA encoding a viral protein. Science 259:1745–1749

Wang B, Ugen KE, Srikantan V, Agadmanyan MG, Dang K, Refaeli Y, Sato AI, Boyer J, Williams WV, Weiner DB (1993a) Gene inoculation generates immune responses against human immunodeficiency virus type 1. Proc Natl Acad Sci USA 90:4156–4160

Wang B, Boyer J, Srikantan V, Coney L, Carrano R, Phan C, Merva M, Dang K, Agadmanyan MG, Gilbert L, Ugen KE, Williams WV, Weiner DB (1993b) DNA inoculation induces neutralizing immune responses against human immunodeficiency virus type 1 in mice and nonhuman primates. DNA and Cell Biology 12:799–805

Wang B, Merva M, Dang K, Ugen KE, Williams WV, Weiner DB (1995a) Immunization by direct DNA inoculation induces rejection of tumor cell challenge. Human Gene Therapy 6:407–418

Wang B, Boyer JD, Ugen KE, Srikantan V, Ayyaroo V, Agadmanyan MG, Williams WV, Newman M, Coney L, Carrano R, Weiner DB (1995b) Nucleic acid-based immunization against HIV-1: induction of protective in vivo immune responses. AIDS 9(suppl A):S159-S170

Wyand MS, Manson KH, Garcia-Moll M, Montefiori D, Desrosiers RC (1996) Vaccine protection by a triple deletion mutant of simian immunodeficiency virus. J Virol 70:3724–3733

Xiang ZQ, Spitalnik S, Tran M, Wunner WH, Cheng J, Ertl HCJ (1994) Vaccination with a plasmid vector carrying the rabies virus glycoprotein gene induces protective immunity against rabies virus. Virology 199:132–140

Yasutomi Y, Reimann KA, Lord CI, Miller MD, Letvin NL (1993) Simian immunodeficiency virus-specific CD8+ lymphocyte response in acutely infected rhesus monkeys. J Virol 67:1707–1711

Yasutomi Y, Robinson HL, Lu S, Mustafa F, Lekutis C, Arthos J, Mullins J, Voss G, Manson K, Wyand M, Letvin NL (1996) Simian immunodeficiency virus-specific cytotoxic T lymphocyte induction through DNA vaccination of rhesus monkeys. J Virol 70:678–681

Zarozinski CC, Fynan EF, Selin LK, Robinson HL, Welsh RM (1995) Protective CTL-dependent immunity and enhanced immunopathology in mice immunized by particle bombardment with DNA encoding an internal virion protein. J Immunol 154:4010–4017

DNA Plasmid Based Vaccination Against the Oncogenic Human T Cell Leukemia Virus Type 1

M.G. AGADJANYAN[1,2], B. WANG[3], S.B. NYLAND[4], D.B. WEINER[1], and K.E. UGEN[4]

1 Introduction

The human T cell leukemia virus type 1 (HTLV-1) holds its place in history as the first human retrovirus established to be associated with disease. Since its discovery, it has been demonstrated to be the causal agent of adult T cell leukemia (ATL) and HTLV-1 associated myelopathy (HAM). In addition, more recently this retrovirus has been suggested to have a role in the etiology and/or pathogenesis of a myriad of autoimmune disorders including rheumatoid arthritis and uveitis (UGEN et al. 1996b). Also, the role of infection with HTLV in accelerating progression to acquired immunodeficiency syndrome (AIDS) in human immunodeficiency virus (HIV)-1 infected individuals has also been suggested (KRAMER et al. 1989). The worldwide seropositivity of HTLV-1 and the related HTLV-2 is estimated to be at

[1]Department of Pathology and Laboratory Medicine, University of Pennsylvania School of Medicine, Philadelphia, PA 19104, USA
[2]Institute of Viral Preparations, Russian Academy of Medical Science, Moscow, Russia 129028
[3]Sinogen Institute, Institute of Zoology, Chinese Academy of Sciences, Beijing, China 100080
[4]Department of Medical Microbiology and Immunology, University of South Florida College of Medicine, Tampa, FL 33612, USA

least 10 million. Coupled with its association to fatal and debilitating disorders as well as its potential role in the pathogenesis of other serious illnesses an effective vaccine is warranted. In this review we will summarize current information on HTLV-1 including immune responses to this virus, correlates of protection as well as potential vaccination strategies including DNA plasmid inoculation.

2 General Biology of Human T Cell Leukemia Virus

The human T lymphotropic virus was first isolated by Robert Gallo's group (POIESZ et al. 1980, 1981). The newly discovered virus, which subsequently was named the human T cell leukemia virus, was the first human retrovirus for which a disease association was made (YOSHIDA et al. 1982). It was therefore designated the human T-cell lymphotropic/leukemic virus type 1, or HTLV-1. Shortly after the discovery of HTLV-1 another human T-cell leukemia virus, named HTLV-2, was isolated from another patient using a similar technique (KALYANARAMAN et al. 1982). An etiologic role for HTLV-1 in ATL and HAM has been demonstrated, but a disease association for HTLV-2 infection has not been established clearly, although some suggestion has been made that it may be associated with a neurodegenerative disorder similar to HAM (CANN et al. 1990; HALL et al. 1996). Both HTLV-1 and HTLV-2 are members of the oncovirinae subfamily of retroviruses. HTLV-1 and HTLV-2 have recently (COFFIN 1996) been placed in a separate group of which the bovine leukemia virus (BLV) is also a member (i.e., the HTLV/BLV subfamily). The discovery and description of HTLV-1 and -2 greatly facilitated the characterization of the etiological agents for AIDS, HIV-1 and -2, which belong to the subfamily of lentiviruses (BARRE-SINOUSSI et al. 1983; GALLO and JAY 1991). Therefore, there are two subfamilies of human retroviruses which infect human CD4-positive T cells but nonetheless produce different and opposite effects. While HIV-1 and 2 results in the death of $CD4^+$ T cells, HTLV-1 infection induces polyclonal and/or monoclonal T cell proliferation.

3 Molecular Biology of Human T Cell Leukemia Virus

The HTLV-1 proviral genome consists of 9032 nucleotides and includes regions coding for structural proteins (*gag* or *env*) and viral protease and polymerase (*pol*), similar to HIV-1. The genome also contains an additional sequence designated pX. This region is adjacent to the envelope (*env*) gene and contains three overlapping regulatory genes encoding the *trans*-activator protein (*tax* p40), transmodulator protein (*rex* p27), and a third gene which encodes for a protein (p21) whose function is unknown (NAGASHIMA et al. 1986; SAGATA et al. 1985; SEIKI et al. 1983). HTLV-1 is most closely related to HTLV-2, with a 66% sequence homology, but

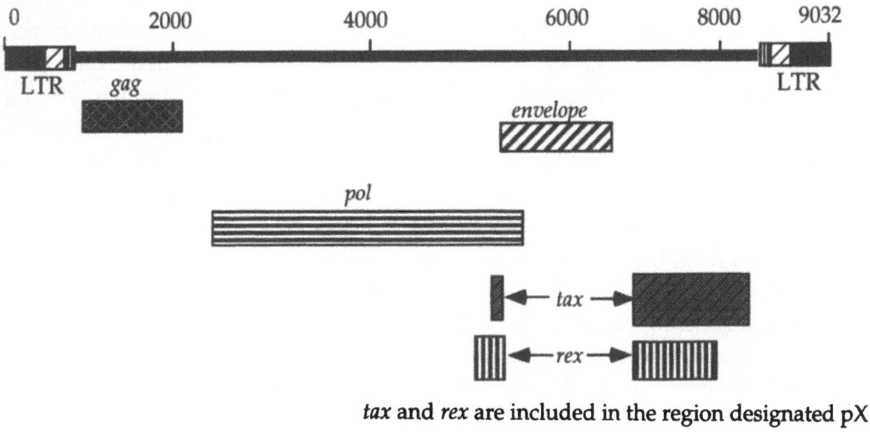

Fig. 1. Genome structure of the human T cell leukemia virus type 1 (HTLV-1). The topographical relationship of the important structural and regulatory genes of the human T cell leukemia virus genome is shown

shares high sequence homologies with additional retroviruses such as BLV and simian T cell leukemia virus. Figure 1 shows the relationship between the various genes of the HTLV-1 genome (YOSHIDA and SEIKI 1987). Unlike other oncoviruses, HTLV-1 does not express an analogue of any known cellular oncogene, nor does it insert its proviral genome into a specific regulatory region of the host DNA. The provirus acts randomly, taking up residence in any open region provided by the host cell's transcriptional activity. Although the mechanisms involved in the oncogenic transformation of infected cells are poorly understood, it is thought that the HTLV-1 *tax* may play an indirect role in transformation by its ability to associate with enhancer binding proteins of the host cell, such as CREB, CREM, NF-kB p50, and SRF (SUZUKI et al. 1993a, b; TANAKA et al. 1990). When *tax* interacts with these proteins, it enables the appropriate cell enhancer to bind, thus causing the up-regulation of transcription (SUZUKI et al. 1994). It has been hypothesized that this interaction involves activation of the interleukin (IL)-2 receptor α-chain, which would link the tax protein activity to the transformation event (INOUE et al. 1986). Another protein hypothesized to play a role in the process of cell transformation is the *rex* protein, which appears to regulate proliferation in a posttranscriptional manner (HIDAKA et al. 1988).

4 Pathogenesis of Human T Cell Leukemia Virus

The ability of HTLV-1 to up-regulate transcription of T cells or to induce their proliferation is linked to two disorders, i.e., ATL and HAM/TSP (tropical spastic pareparesis) (GESSAIN et al. 1985; SMITH and GREENE 1991). HAM/TSP is a neurodgenerative disorder which mimics some of the clinical manifestations of multiple sclerosis (GESSAIN and GOUT 1992). HAM/TSP can develop within several months of infection by HTLV-1. In this disorder infected cells in the CNS present antigen, eventually leading to a chronic inflammatory state. The ensuing tissue damage is presented clinically as a progressive demyelinating syndrome with weakness starting at the extremities, and may resolve to varying degrees over time. In contrast to HAM/TSP, the second disorder, ATL, is attributed to the monoclonal proliferation of infected cells and presents abruptly during later adulthood with clinical latency lasting up to 50 years (KAPLAN and KHABBAZ 1993). ATL manifests clinically as a leukemic infiltration of a number of organ systems accompanied by moderate to profound immunosuppression (BUNN et al. 1983). The leukemia is frequently aggressive and generally untreatable, and death usually occurs within several months after diagnosis. Epidemiological studies of the two disorders have revealed factors suspected of influencing the direction of an HTLV-1 infection. These factors include the type and magnitude of the immune responses to HTLV-1 antigens, organ tropism, and the age at which an individual acquires an infection (RENJIFO et al. 1996). HAM/TSP is associated with infections acquired during early adulthood, while the emergence of ATL corresponds to a history of virus exposure from the prenatal period up through the first year of life, and breastfeeding appears to be an important epidemiological factor (HINO et al. 1996). In addition, certain geographical regions contain distinct pockets of high ATL and HAM/TSP incidence in conjunction with endemic HTLV-1 infection. These clusters are most prevalent in the Caribbean Basin (BLATTNER et al. 1982), Africa, and southwestern Japan (HINUMA et al. 1982; TAJIMA et al. 1982). Nagasaki has been shown to be especially vulnerable to HTLV-1 infection, with the prevalence of carriers being 10% in older persons and 4% in pregnant women (HINO et al. 1985). Overall, at least 10 million people are infected with HTLV-1/2 worldwide (DE THE and BOMFORD 1993). Individuals infected with HTLV-1 develop antibody responses against the virus and become lifelong carriers. It is estimated that one of every thousand infected individuals per year will become afflicted with ATL, with a lifetime risk of developing ATL being 2.5%–5%. The lengthy latency period has made the pathogenesis of HTLV-1 infection and its associated disease manifestations somewhat difficult to understand.

Transmission routes are better understood and have been established to be sexual spread, parenteral injection of contaminated blood, and mother-to-child (vertical) transmission. The evidence for male to female transmission of HTLV-1 is suggested by the prevalence of elderly female carriers (TAKATSUKI et al. 1982) who are wives of seropositive males and by the presence of HTLV-1 infected T cells in the infected husbands' seminal fluids (NAKANO et al. 1984). However, because this higher incidence of female carriers is noted only in female populations over 60 years of age, male to female sexual transmission of HTLV-1 is thought to be inefficient. Nearly 50% of patients receiving a blood transfusion from an HTLV-1 infected source also become carriers of infection (HINO et al. 1985; OKOCHI et al. 1985). The most striking aspect of HTLV-1 transmission by blood transfusion is its reliance on cell to cell association, i.e., frozen cell-free plasma is not infectious. This property is consistent with tissue culture studies demonstrating that infection by HTLV-1 virions is much less efficient when cell-free viral suspensions are used rather than contact cell cultures

(Agadjanyan et al. 1994a; Clapham et al. 1983; DeRossi et al. 1985; Fan et al. 1992; Miyoshi et al. 1981; Weiss et al. 1985). Mother-to-child transmission of infection was first observed epidemiologically from the association of carrier clusters in families with the hallmark of a maternal seroconversion followed by HTLV-1 infection of children born after the conversion took place, while children born prior to the maternal seroconversion remained uninfected (Tajima et al. 1982). It was noted that in these clusters HTLV-1 infections occurred at a predominately early stage of life, which is compatible with the notion and importance of maternal-infant transmission. The genetic route for HTLV-1 infection was eliminated as a possibility when it was found that the virus is not endogenous to humans (Yoshida et al. 1982). Although a case is occasionally reported in which HTLV-1 infected cells are identified in the cord blood of infants born to infected mothers (Komuro et al. 1983), this pre- and perinatal form of transmission does not appear to play a major role in the spread of the virus. Since the lag time for the development of antibodies is approximately 2 months from the time of infection by blood transfusion (Okochi et al. 1985) or by oral transmission (Yamanouchi et al. 1985), seroconversions from a perinatally acquired infection would be expected to take approximately 2–3 months to develop. Instead, seroconversions took at least 6 months to be detected and the majority of conversions did not occur until 12 months after birth (Hino et al. 1985) suggesting that vertical transmission of the virus could be attributed to a postnatal route, as in breast feeding. ATL prevention programs in Japan advise HTLV-1 infected mothers to avoid breastfeeding their infants. Although the program has helped to reduce the spread of HTLV-1 infection to some extent, it has proven impractical for other HTLV-1 endemic areas, such as the Caribbean basin countries where economic pressures preclude the use of expensive bottle-feeding equipment and formulas, and where the medical and educational resources of Japan are unavailable. More accessible forms of prevention, including an effective vaccine strategy, would better serve such communities. For this reason, the development of an HTLV-1 vaccine has become the focus of several research groups.

As indicated earlier it is estimated that at least 10 million individuals worldwide are infected with HTLV-1. This is certainly a sufficiently high enough incidence of infection to warrant the development of an effective vaccine. Coupled with the association of HTLV-1 infection with serious disorders as well as its potential role as a factor for the development of AIDS in HIV-1/HTLV-1 co-infected individuals an effective vaccine against HTLV-1 would be a significant development. In addition, strategies used to develop a vaccine against HTLV-1 may be applicable in the effort to develop an efficacious vaccine against HIV-1.

5 Vaccine Development

Currently accepted vaccine strategies generally employ two resources for producing immunogenic material – live infectious agents, or inactivated/subunit preparations.

The polio and smallpox vaccines are examples of the use of live, attenuated agents to stimulate protective T_{helper} (Th) and $T_{cytotoxic}$ (cytotoxic T lymphocyte, CTL) responses. This occurs during a nonpathogenic infection of the host and leads to an additional humoral immune response. The use of this type of vaccine for protection against human retroviruses still awaits the development of a nonpathogenic, yet immunogenic strain; an unlikely possibility for the near future. Subunit preparations include the recently developed hepatitis B surface antigen vaccine, which has been reported to stimulate protective T_{helper} and humoral immune responses. Although subunit vaccines appear to work with some pathogens, their effectiveness with other pathogens has not been confirmed. An additional strategy includes the vaccinia virus, which has been engineered to operate as an immunogenic expression vector and can elicit broader T cell immune responses (Moss 1988). It was initially believed that the vaccinia system would prove to offer a safe and simplified alternative to the other vaccination schemes currently available. However, it was later found that the viral vector itself presents antigens which can react with the immune system of the vaccinated individual, thus leading to a type of hypersensitivity reaction. This reaction renders the vaccinia construct impotent after a limited number of uses, as the construct would be neutralized before the desired immunogen could be processed.

The ideal vaccine for human retroviruses (i.e., HTLV-1 and HTLV-2) would act as an effective subunit or attenuated system, eliciting the whole spectrum of protective immune responses in the vaccinated subject without the risk of reversion to a pathogenic wild-type expression and without the risk of deleterious immune responses to the delivery construct. To this end, the DNA plasmid vaccination technology has been under investigation (Ulmer et al. 1993; Wang et al. 1993). This methodology involves the cloning of DNA sequences into an appropriate eukaryotic expression vector followed by delivery to the host through a number of different techniques. Genes cloned in this manner can be manipulated so that single proteins, or an extended genome excluding the genes that might lead to pathogenesis, can be presented.

The potential efficacy of the DNA plasmid vaccination system against human retrovirus (HIV-1) was also shown. We have been successful in eliciting neutralizing antibodies as well as CTL activity in rodents and nonhuman primates with HIV-1 envelope expressing constructs (Boyer et al. 1996; Ugen et al. 1996a, 1997; Wang et al. 1993, 1995). Importantly, this methodology has been successful in decreasing viral load in HIV-1 infected chimpanzees as well as protecting naive chimpanzees from heterologous challenge (Boyer et. al., 1997). This methodology can, in principle, be applied to other infectious agents such as HTLV-1.

5.1 Immune Response to HTLV-1

In order to design a vaccine for HTLV-1, it is necessary to document and understand the immune responses elicited against this retrovirus. Like its retroviral relatives, HIV-1 and -2, the envelope glycoproteins of HTLV-1 can act as major antigens which are recognized by serum antibodies in infected individuals. The HTLV-1 transmembrane glycoprotein gp21 serves to anchor the exterior glycoprotein gp46 by a noncovalent association and assists in the attachment of gp46 to the host cell membrane (BOHNLEIN et al. 1991). Both gp46 and gp21 are derived from the proteolytic cleavage of an envelope precursor, gp66, which is encoded by the viral *env* gene (BOLOGNESI et al. 1978). Epitope mapping studies with the HTLV-1 envelope glycoprotein have not been as extensive as similar studies involving HIV-1, nonetheless several immunodominant B cell epitopes and a T cell epitope have been identified to date. A major neutralizing linear determinant has been identified in the external glycoprotein which spans amino acids 187–196. (DELAMARRE et al. 1994). Mutagenesis of this region has indicated its role in cell to cell fusion. Other regions within the external glycoprotein identified as of potential functional importance are amino acids 53–75 and 90–98 (DELAMARRE et al. 1994; PALKER et al. 1992) as well as two regions in the COOH-terminal amino acids 213–236 and 287–311 (BABA et al. 1993; LAIRMORE et al. 1992; LAL et al. 1991). One region in the gp21 transmembrane glycoprotein has been suggested to elicit neutralizing antibodies (i.e., amino acids 346–368) (DELAMARRE et al. 1994) in the central region of the molecule (amino acids 191–214) and another near the COOH-terminal (amino acids 242–257) (LAIRMORE et al. 1992; LAL et al. 1991). The external envelope glycoprotein gp46 also contains a region (amino acids 190–209) which was shown to be a target for cytotoxic T cell activity (PALKER et al. 1990).

HTLV-1 appears to be a good candidate for vaccine development because of the low variability within the envelope glycoprotein and the ability of this retrovirus to infect small animals; i.e., rats and rabbits (AMI et al. 1992; IBRAHIM et al. 1994; ISHIGURO et al. 1992; MIYOSHI et al. 1985). The hypervariability noted in HIV-1 can reduce the ability of neutralizing antibodies to appropriately interact with the immunogenic glycoprotein, thus reducing the efficacy of a vaccine designed against gp120 and this region in particular. With HIV-1, up to 27% of the envelope amino acid sequences may vary between different isolates (COFFIN 1986). Fortunately, this problem does not exist to any significant degree with HTLV-1. Envelope glycoprotein sequences from different HTLV-1 isolates were compared and demonstrated only a 2% overall variance in amino acid sequences, thus a high level of conservation was indicated for the HTLV-1 *env* sequence (GRAY et al. 1990). It is unlikely that a shift in variability will occur since mutations introduced into the HTLV-1 *env* gene results in a nonfunctional protein which is unable to form syncytia. When similar mutations are introduced into the HIV-1 genome, envelope function continues. This property of HTLV-1 suggests that it may contain the appropriate target material for the development of a broadly neutralizing vaccine based on the envelope glycoprotein.

In addition to the low variability in the envelope glycoprotein the important ability of anti-envelope antibodies in patient sera to cross-neutralize HTLV-1 strains originating from disparate geographic regions around the world suggests that a vaccine developed against one strain may be effective in neutralizing other strains (CLAPHAM et al. 1984; HOSHINO et al. 1985). Also, it appears that natural immunity to HTLV-1 can be effective since maternal antibodies are able to protect newborns against congenital transmission including that which occurs by the epidemiologically important breast feeding route (TAKAHASHI et al. 1991).

5.2 Anti-HTLV Envelope Glycoprotein Responses by Conventional Vaccines

Other work on the demonstration of functional antibody responses involved the use of a vesicular stomatitis virus pseudotype displaying the HTLV-1 envelope. This preparation was used to demonstrate the neutralizing ability of antibodies specific for the envelope glycoproteins (CLAPHAM et al. 1984; HOSHINO et al. 1985). Experimental vaccination with these viral glycoproteins were also able to inhibit the fusion activity of whole HTLV-1 (HOSHINO et al. 1983; NAGY et al. 1983). Likewise, cynomologous monkeys were protected from HTLV-1 challenge when immunized with viral envelope components which were produced by genetically engineered bacteria (NACAMURA et al. 1987). In addition, viral envelope gene products are recognized by HTLV-1 seropositive patients (KANNER et al. 1986) and experimental animals treated with these reagents can produce neutralizing antibodies (KIYOKAWA et al. 1984; SAMUEL et al. 1984). Most neutralizing antibodies against retroviruses, including HIV-1 and HIV-2, are directed toward the envelope glycoprotein. In the case of HTLV-1, an attractive target for vaccine development can be found in this region of the virus.

Other strategies which have been utilized in HTLV-1 vaccine development include soluble HTLV-1 viral protein (DEZUTTI et al. 1990), synthetic peptides (TANAKA et al. 1994), recombinant vaccinia (FRANCHINI et al. 1995) and recombinant adenovirus (KAZANJI et.al., in press). To some degree, all of these preparations resulted in protection of experimental animal models (i.e., monkeys or rabbits) from infection by HTLV-1.

5.3 Anti-HTLV Envelope Glycoprotein Responses by DNA Plasmid Vaccines

5.3.1 Humoral Immune Responses

Using our studies on HIV-1 as a guide we have embarked on work which targets the HTLV-1 envelope glycoprotein for a DNA plasmid based vaccine. For DNA plasmid inoculation studies in rabbits we used a construct designated pcTSP/AT-K.env (REDDY et al. 1988) which contains the gene for gp46 + gp21 (i.e., external and transmembrane glycoproteins) from HTLV-1 (Fig. 2A). For rat studies we used a dual vector system, i.e., pgTAXLTR, which contains the genes for HTLV-1 *tax*, and *env* and pc*REX*,which contains the gene for HTLV-1 *rex* (RIMSKY et al. 1988) (Fig. 2B, C). The coding region for HTLV-1 *rex* and tax genes were included in the vector cocktail for the generation of immune responses against these important viral proteins. Therefore these constructs following experimental mutation

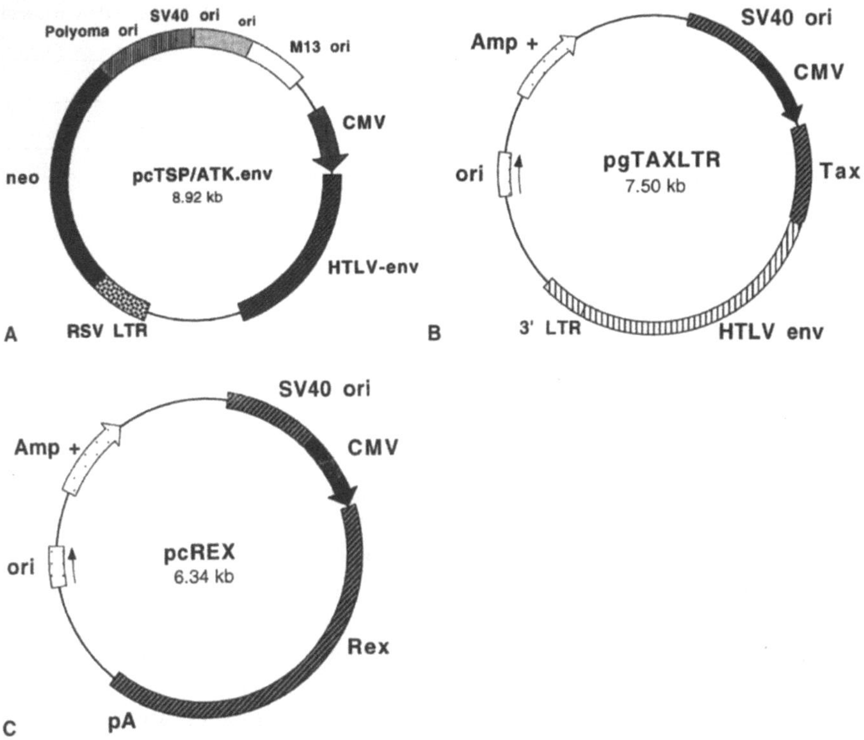

Fig. 2A–C. Human T cell leukemia virus type 1 (HTLV-1) envelope glycoprotein expressing DNA plasmids used in the study: **A** the pcTSP/ATK.env plasmid used in the rabbit inoculation experiments; **B** the dual pgTAXLTR and **C** pc *REX* plasmids used in the rat inoculation experiments

of the *rex* and tax genes could be in theory used as a therapeutic vaccine. Further analysis of immune responses against these viral antigens are planned. Possible up-regulation of the HTLV env protein synthesis by *rex* gene (Hidaka et al. 1988) was another reason for including this gene into the vector.

Humoral immune responses were detected after single or multiple inoculations of 100 μg of the constructs after pretreatment with bupivacaine in both New Zealand White rabbits (Agadjanyan et al. 1994b) as well as Fisher 344 rats (Ugen et al. 1995) (Table 1). Specifically immune responses in the experimental animals were evident, as measured by ELISA against several recombinant HTLV-1 envelope glycoproteins (i.e., RE-3: amino acids 165–306 and RE-6: amino acids 165–440) purchased from Repligen Corporation (Cambridge, MA). RE-3 contains the COOH-terminal half of gp46 while RE-6 contains the COOH-terminal half of gp46 + 75% of gp21. Both recombinant proteins contain the major neutralizing B cell epitopes of the glycoprotein as well as the T cell epitope. In addition, humoral immune responses were also measured against two peptides from gp46, designated Env-1 (amino acids 191-214) and Env-5 (amino acids 242–257) (Rudolph and Lal 1993). Also, immune responses in rabbits and rats to another described neutralizing epitope, amino acids 272–292, were noted in our study.

Table 1. Immune responses in experimental animals used as targets for HTLV-1 *env* DNA plasmid inoculation

DNA vaccine	Rabbit					Rat				
	1	2	3	4	5	1	2	3	4	5
HTLV pcTSP/ATK	+	+	+	+	+	nt	nt	nt	nt	nt
HTLV pcrex + pctax	nt	nt	nt	nt	nt	+	+	+	+	+

1, RE-3; 2, RE-6; 3, Env-1; 4, Env-5; 5, peptide 6; RE-3, amino acids 165–306; RE-6, amino acids 165–440; Env-1, amino acids 191–214 (LPHSNLDHILEPSIPWKSKLLTLV); Env-2, amino acids 242–257 (SPNVSVPSSSSTPLLY); peptide 6, amino acids 272–292 (NWTHCFDPQIQAIVSSPCHNS).

5.3.2 Neutralizing Antibody Responses

Anti-syncytial activity of antisera from rabbits and rats inoculated with an HTLV-1 envelope expressing constructs were measured by standard procedures (Agadjanyan et al. 1994a; Clapham et al. 1984). Rabbits were inoculated three times with 100 µg of an HTLV-1 envelope expressing construct (pcTSP/ATK.env) after pretreatment with bupivacaine (Fig. 3A). Fifteen weeks after the final inoculation animals were bled and the ability of the antisera to inhibit syncytia between HTLV-1 infected (MT-2) cells and a target B cell line (BJAB-WH) was tested. Sera were mixed with both cell lines and 48 h later syncytia formation was quantitated. Rabbit 1 was inoculated with a control plasmid which did not express the HTLV-1 envelope glycoprotein. Significant inhibition of syncytia was noted in all three of the rabbits which were inoculated with the HTLV-1 envelope expressing plasmid while the control rabbit failed to inhibit syncytia formation (Agadjanyan et al. 1994b). Likewise, Fisher 344 rats were inoculated with either (a) 100 µg of an HTLV-1 envelope expressing constructs (i.e., pgTAXLTR + pc*REX*) after pretreatment with bupivacaine, (b) 5 µg of a recombinant HTLV-1 protein (RE-6) or (c) a control DNA plasmid not expressing the HTLV-1 envelope construct (Fig. 3B). Twelve days later the rats were bled, sera collected and the ability to inhibit syncytia formation was performed using the cell lines and conditions described above. Inhibition data is shown for pre- and post-inoculation bleeds. Rats inoculated with either the experimental DNA plasmid or the recombinant protein demonstrated an ability to inhibit syncytia formation. In fact, DNA plasmid inoculation appeared to produce a more potent neutralization as evidenced by the higher percent inhibition of syncytia at a serum dilution of 1:27 when compared to serum from rats vaccinated with recombinant protein (Ugen et al. 1995).

5.3.3 Cellular Immune Responses

T cell proliferative responses in rabbits inoculated with an HTLV-1 envelope expressing DNA plasmid were measured. The rabbits were inoculated intramuscularly with 100 µg of the pcTSP/ATK.env plasmid 24 h after pretreatment with

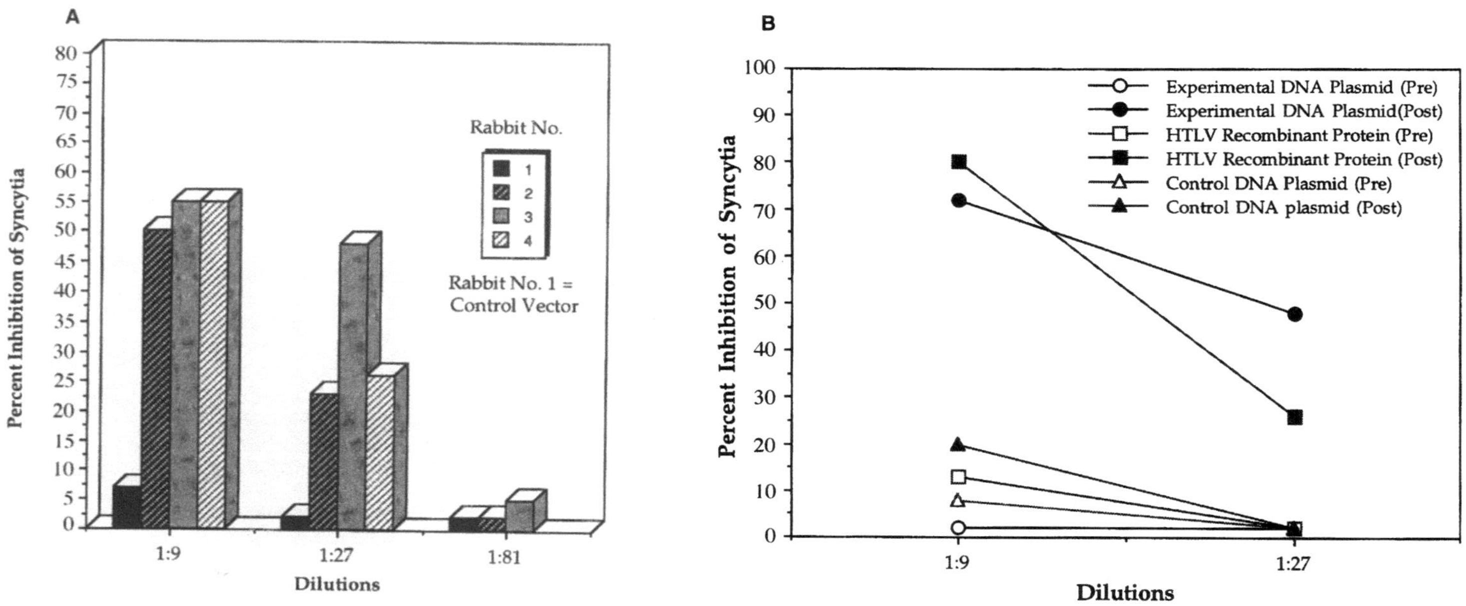

Fig. 3A, B. Anti-syncytial activity of antisera from rabbits and rats inoculated with an HTLV-1 envelope expressing construct. **A** Rabbits 1–3 were inoculated with the pcTSP/ATK.env plasmid while rabbit 4 received a control plasmid with the specific HTLV-1 gene sequence deleted. **B** Rats were inoculated with the pgTAXLTR and pc*REX* plasmids, an HTLV-1 recombinant envelope glycoprotein plasmid (RE-6) or a control plasmid which had the HTLV-1 envelope gene deleted. *Pre* refers to a pre-inoculation measurement and *post* refers to a post-inoculation measurement. Details of the methods are included in the text

0.5% bupivacaine. Blood was subsequently collected from the rabbits after vaccination and peripheral blood mononuclear cells (PBMCs) were isolated and stimulated with either a positive control (PHA, phytohemagglutinin, at 4 µg/ml) or a specific HTLV-1 envelope recombinant protein (RE-6) at 1 µg/ml. Rabbit 1 was inoculated with a control plasmid which did not express the HTLV-1 envelope glycoprotein. Specific proliferation was measured by incorporation of [^{3}H]thymidine (AGADJANYAN et al. 1994b). The results indicate that specific proliferation occurred in the rabbits inoculated with the HTLV-1 *env* expressing plasmid (Fig. 4).

5.3.4 Protection from HTLV-1 Infection by DNA Plasmid Based Vaccine

The ultimate test of the efficacy of a vaccine is its ability to prevent infection and/or disease. To this end Fisher rats which were inoculated with an HTLV expressing construct and which demonstrated significant anti-envelope glycoprotein antibodies were challenged with HTLV-1 infected MT-2 cells. The protocol for this viral challenge is as follows:

Fisher 344 rats were vaccinated three times (following pre-treatment with 0.5% bupivacaine-HCl) with 100 µg of the HTLV-1 DNA plasmids (pgTAXLTR +

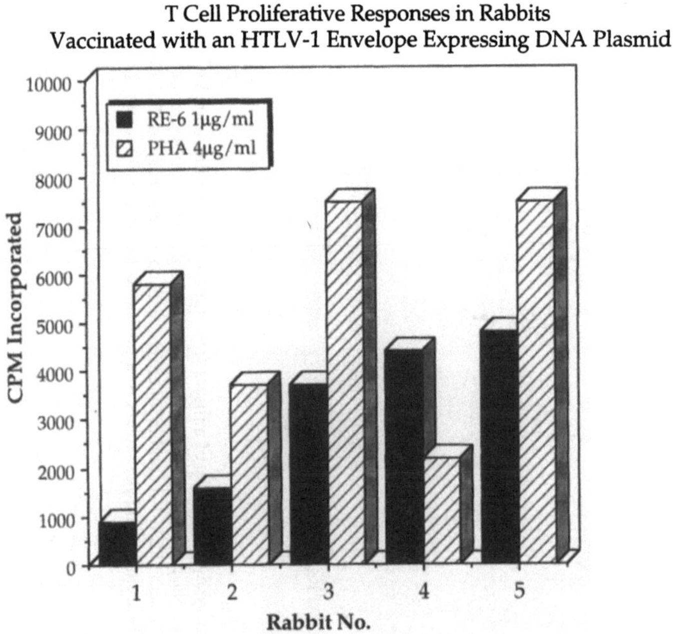

Fig. 4. T cell proliferative responses in rabbits inoculated with an HTLV-1 envelope expressing DNA plasmid. Rabbit 1 was inoculated with a control DNA plasmid in which the HTLV-1 envelope gene was deleted. Rabbits 2–5 were inoculated with the HTLV-1 envelope expressing the pcTSP/ATK.env plasmid. Details of the assay are included in the text

pc*REX*) at 2 week intervals. In addition other rats were inoculated with a control plasmid according to the same schedule. Humoral immune responses against an HTLV-1 envelope glycoprotein were measured in the rats inoculated with the HTLV *env* plasmid. Two weeks following the final inoculation all of the rats were challenged with 3×10^6 HTLV-1 infected human MT-2 cells by intraperitoneal injection. Twelve weeks after challenge with MT2 cells the rats were bled and DNA was prepared from the PBMCs and subjected to PCR using the following primers from the pX gene region (no. 1 = nt 7336–7355: CGGATACCCAGTCTACG-TGT and no. 2 = nt 7473–7492: CT*GAG*CCGATAACGCGTCCA). Figure 5 shows the results of these analyses. The gel lane designations for the figure are as follows: lane 1, DNA size markers; lane 2, DNA from HTLV-1 infected MT-2 cells; lane 3, DNA from one of the MT-2 challenged rats vaccinated with the HTLV-1 env DNA plasmid in which actin specific primers were used in the reaction as a control; lanes 4 and 5, DNA from rats inoculated with control plasmid; lanes 6 and 7, DNA from rats inoculated with experimental HTLV-1 *env* expressing plasmid. These results demonstrated that the vaccinated rats had the px PCR product absent from the DNA of their PBMCs. The positive actin band (lane 3) suggested that DNA was present in these samples and the lack of px signal (lanes 6, 7) was not due to an absence or low concentration of DNA.

Recently KAZANJI et al. (in press) have also reported partial protection from infection by cell associated HTLV-1 in rats vaccinated with an HTLV-1 env expressing construct. In this study they found anti-envelope CTL activity, but not humoral immune responses against HTLV-1 env. Based upon these findings they concluded that CTL activity elicited by the DNA plasmid inoculation technique was correlated with protection. In our study CTL activity was not measured, however, humoral neutralizing antibodies were present in the rats, which were

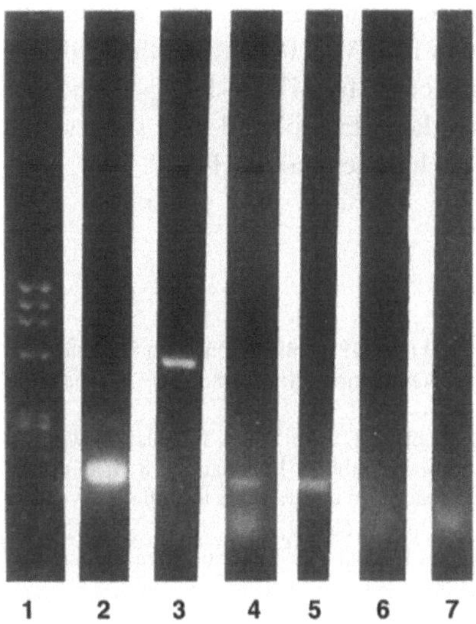

Fig. 5. Protection of Fisher 344 rats from infection with HTLV-1 MT-2 cells by HTLV-1 envelope glycoprotein DNA plasmid inoculation. Details and a description of the analysis are included in the text

protected after challenge. It is clear than additional studies are necessary to establish the relative protective roles of cellular and humoral immunity against HTLV-1 infection and disease.

6 Discussion and Summary

Successful vaccine development for human retroviral infections are of considerable importance due to the worldwide AIDS pandemic. However, the first human retrovirus described which produces disease, HTLV-1, also infects at least 10 million people worldwide and results in fatal and debilitating disorders. Therefore, an effective vaccine for this retrovirus is also warranted. Several characteristics of HTLV-1 suggest that a vaccine based on the envelope glycoprotein may be feasible. These include: (a) minimal hypervariability of the envelope glycoprotein (unlike HIV-1), indicating that a vaccine may be effective against all viral isolates and (b) the ability of envelope glycoprotein vaccinogens to protect monkeys from challenge with cell associated HTLV-1. Also, some evidence suggests that humoral immune responses alone may be sufficient to protect against infection. Cellular immune responses, however, may also play some role in protection. Therefore, an HTLV-1 envelope glycoprotein expressing DNA plasmid vaccine may be useful in eliciting protective immune responses against this retrovirus. This review has summarized the current knowledge concerning HTLV-1 biology and pathogenesis as well as the immune responses elicited against this retrovirus during natural infection and experimental vaccination. We have summarized our work to date on the HTLV-1 envelope DNA plasmid vaccines in the rabbit and rat infection models for this retrovirus. Humoral and cellular immune responses in the animal systems have been demonstrated. Importantly, current studies indicate that this vaccination methodology can protect rats from infection with HTLV-1. Recently, other authors have also reported protection of rats from infection by HTLV-1 by vaccination with envelope expressing DNA constructs. All the data reported here suggest the potential utility of the DNA plasmid vaccine technique against HTLV.

References

Agadjanyan M, Ugen K, Wang B, Williams W, and Weiner D (1994a) Identification of an 80-kilodalton membrane glycoprotein important for human T-cell leukemia virus type I and type II syncytium formation and infection. J Virol 68:485–493

Agadjanyan M, Wang B, Ugen K, Villafana T, Merva M, et al. (1994b) DNA inoculation with an HTLV-I envelope DNA construct elicits immune responses in rabbits. In: Chanock RM, Ginsberg HS, Brown F, Lerner RA (eds) Vaccines 94: Modern approaches to new vaccines including prevention of AIDS. Cold Spring Harbor, Cold Spring Harbor, NY, pp 47–53

Ami Y, Kushida S, Matsumura M, Yoshida Y, Kameyama T, et al. (1992) Vertical transmission of HTLV-I in HTLV-I carrier rat. Jpn J Cancer Res 83:1241–1243

Baba E, Nakamura M, Tanaka Y (1993) Multiple neutralizing B-cell epitopes recognized by human monoclonal antibodies. J Immunol 151:1013–1024

Barre-Sinoussi F, Cherman JC, Rey F, Nugeyre MT, Chameret S, et al. (1983) Isolation of a T-lymphotropic retrovirus from a patient at risk for acquired immune deficiency syndrome (AIDS). Science 220:868–871
Blattner WA, Kalyanaraman VS, Robert-Guroff M, Lister TA, Galton AG, et al. (1982) The human C retrovirus: HTLV in blacks from the Caribbean region, and relationship to adult T cell leukemia/lymphoma. Int J Cancer 30:257
Bohnlein E, Berger J, Hauber J (1991) Functional mapping of the human immunodeficiency virus type 1 rev RNA binding domain: new insights into the domain structure of rev and rex. J Virol 65:7051–7055
Bolognesi D, Montelaro R, Frank H, Schafer W (1978) Assembly of type C oncornaviruses: a model. Science 199:183–186
Boyer JD, Ugen KE, Wang B, Agadjanyan MG, Bagarazzi ML, et al. (in press) Journal of Infectious Diseases DNA vaccination as an anti-HIV immunotherapy in infected chimpanzees
Boyer JD, Ugen KE, Wang B, Agadjanyan MG, Gilbert L, et al. (1997) Protection from high dose heterologous HIV-1 challenge in chimpanzees by DNA vaccination. Nature Medicine 3:526–532
Boyer JD, Wang B, Ugen K, Agadjanyan M, Javadian A, et al. (1996) Protective anti-HIV immune responses in non-human primates through DNA immunization. J Med Primatol 25:242–250
Bunn P, Schecter G, Jaffe E (1983) Clinical course of retrovirus-associated adult T cell lymphoma in the United States. NEJM 309:257–264
Cann AJ, Chen ISY (1990) Human T-cell leukemia virus types I and II. In: Fields BN, Knipe DM, Howley PM (eds) Virology, Raven , NY, pp 1849–1880
Clapham P, Nagy K, Cheingsong-Popov R, Exley M, Weiss RA (1983) Productive infection and cell-free transmission of human T-cell leukemia virus in a nonlymphoid cell line. Science 222:1125–1127
Clapham P, Nagy K, Weiss R (1984) Pseudotypes of human T-cell leukemia virus types 1 and 2: neutraliztion by patient's sera. Proc Natl Acad Sci USA 81:2886–2890
Coffin J (1986) Genetic variation in AIDS virus. Cell 46:1–4
Coffin JM (1996) Retroviridae: the viruses and their replication. In: BN Fields, DM Knipe, PM Howley (eds) Virology Lippincott-Raven, Philadelphia, pp 1767–1880
De The G and Bomford R (1993) An HTLV-I vaccine: Why, how, whom? AIDS Res Hum Retroviruses 9:381–386
Delamarre L, Pique C, Pham D, Tursz T, Dokhelar M-C (1994) Identification of functional regions in the human T-cell leukemia virus type I SU glycoprotein. J Virol 68:3544–3549
DeRossi A, Aldovani A, Franchini G, Mann D, Gallo RC (1985) Clonal selection of T-lymphocytes infected by cell-free human T-cell leukemia/lymphoma virus type I: parameters of virus integration and expression. Virology 143:640–645
Dezutti CS, Frazier FE, Huffl Y, Stromberg PC, Olsen RG (1990) Subunit vaccine protects Macaca nemestrina (pig-tailed macaque) against simian T-cell lymphotropic virus type 1 challenge. Cancer Res 50:5687–5691
Fan N Gavalchin J, Paul B, Wells K, Lane MJ, et al. (1992) Infection of peripheral blood mononuclear cells and cell lines by cell free human T-cell lymphoma/leukemia virus type I. J Clin Microbiol 30: 905–910
Franchini G, Tartaglia J, Markham P, Benson J, Fullen J, et al. (1995) Highly attenuated HTLV type I env poxvirus vaccines induce protection against a cell – associated HTLV type I challenge in rabbits. AIDS Res Hum Retroviruses 11:307–313
Gallo RC, Jay G (1991) The human retroviruses. In: Gallo RC, Jay G (eds) The human retroviruses. Academic, NY, p 421
Gessain A, Barin F, Vernant JC, Gout O, Maurs L, et al. (1985) Antibodies to human T-lymphotropic virus type-I in patients with tropical spastic paraparesis. Lancet ii: 407–410
Gessain A, Gout O (1992) Chronic myelopathy associated with human T lymphotropic virus type 1 (HTLV-1). Ann Intern Med 117:933–946
Gray GS, White M, Bartman T, Mann D (1990) Envelope gene sequence of HTLV-1 isolate MT-2 and its comparison with other HTLV-1 isolates. Virology 177:391–395
Hall WW, Ishak R, Zhu SW, Novoa, P Eiraku N, et al. (1996) Human T lymphotropic virus type II (HTLV-II): epidemiology , molecular properties, and clinical features of infection. J Acquir Immune Defic Syndromes Hum Retrovirol 13:S204-S214
Hidaka M, Inoue J, Yoishida M, Seiki M (1988) Post-transcriptional regulator (rex) of HTLV-1 initiates expression of virla structural proteins but suppresses expression of regulatory proteins. EMBO J72:519–523
Hino S, Katamine S, Miyata H, Tsuji Y, Yamabe T, et al. (1996) Primary prevention of HTLV-I in Japan. J Acquir Immune Defic Syndromes Hum Retrovirol 13:S199-S203

Hino S, Yamaguichi K, Katamine S (1985) Mother-to-child transmission of human T-cell leukemia virus type I. Jpn J Cancer Research 76:474–480
Hinuma Y, Komoda H, Chosa T, Kondo T, Kohakura M, et al. (1982) Antibodies to adult T cell leukemia virus-associated antigen (ATLA) in sera from patients with ATL and controls in Japan: a nation-wide seroepidemiologic study. Int J Cancer 29:631–635
Hoshino H, Shinoyama M, Miwa M and Sugimura T (1983) Detection of lymphocytes producing a human retrovirus associated with adult T-cell leukemia by syncytium induction assay. Proc Natl Acad Sci USA 80:7337–7341
Hoshino H , Weiss R, Miyoshi I, Yoshida M, et al. (1985) Human T-cell leukemia virus type-I: psedotype neutralization of Japanese and American isolates with human and rabbit sera. Int J Cancer 36: 671–675
Ibrahim F, Flette L, Gessain ANB, De The G, et al. (1994) Infection of rats with human T-cell leukemia virus type 1: susceptibility of inbred strains, antibody response and provirus location. Int J Cancer 58:446–451
Inoue J, Seiki M, Taniguchi T (1986) Induction of interleukin 2 receptor gene expression of by p40x encoded by human T-cell leukemia virus type 1. EMBO J5:2883–2888
Ishiguro N, Abe M, Seto K, Sakurai H, Ikeda H, et al. (1992) A rat model of human T lymphocyte virus type I (HTLV-I) infection. I Humoral antibody response, provirus integration and HTLV-I associated myelopathy/tropical spastic paraparesis-like myelopathy in seronegative HTLV-I carrier rats. J Exp Med 176:981–989
Kalyanaraman VC, Sarngadharan MG, Robert-Guroff M, Miyoshi I, Blayney D, et al. (1982) A new subtype of human T-cell leukemia virus (HTLV-II) associated with a T-cell variant of hairy cell leukemia. Science 218:571–573
Kanner S, Cheng-Mayer C, Geffen R (1986) Human retroviral env and gag polypeptides: serologic assays to measure infection. J Immunol 137:674–678
Kaplan JE, Khabbaz RF (1993) The epidemiology of human T-lymphotropic viruses types I and II. Med Virol 3:137–148
Kazanji M, De The G, Bessereau JL, Schulz T, Shida H, et al. (1997) Vaccination of rats with recombinant adenovirus 5, DNA plasmids and vaccinia viruses containing the HTLV-1 env gene. AIDS Res Hum Retroviruses (in press)
Kiyokawa T, Yoshikura H, Hattori S, Seiki M, Yoshida M (1984) Envelope proteins of human T-cell leukemia virus. Expression in Escherichia coli and its applications to studies of env-gene functions. Proc Natl Acad Sci USA 81:6202–6206
Komuro A, Hayami M, Fujii H, Miyahara M (1983) Vertical transmission of adult T-cell leukemia virus. Lancet1:240
Kramer A (1989) Spontaneous lymphocyte proliferation in symptom free HTLV-1 positive Jamaicans. Lancet 2:923–924
Lairmore MD, Rudolph DL, Roberts BD, Dezzutti CS, Lal RB (1992) Characterization of a B-cell immunodominant epitope of human T-lymphotropic virus type 1 (HTLV-I) envelope gp46. Cancer Lett 66:11–20
Lal R, Rudolph D, Lairmore M, Khabbaz R, Garfield M, et al. (1991) Serologic discrimination of human T cell lymphotropic virus infection by using a synthetic peptide-based enzyme immunoassay. J Infect Dis 163:41–46
Miyoshi I, Kubonishi I, Yoshimoto S, et al. (1981) Type C particles in a cord T cell derived by cocultivating normal human cord leukocytes and human leukemic T cells. Nature 294:770–771
Miyoshi I, Yoshimoto S, Kubonishi I, Fujishita M, Ohtsuki Y, et al. (1985) Infectious transmission of human T-cell leukemia virus to rabbits. Int J Cancer 35:81–85
Moss B (1988) Roles of vaccinia virus in the development of new vaccines. Vaccine6:161–163
Nacamura H, Hayami M, Ohta Y, Ishikawa K, Tsuimoto H, et al. (1987) Protection of cynomologus mankeys agaist infection by human T-cell leukemia virus type-I by immunization with viral env gene products produced in Escherichia coli. Int J Cancer 40:403–407
Nagashima K, Yoshida M and Seiki M (1986) A single species of pX mRNA of human T-cell leukemia virus type I encodes trans-activator p40x and two other phosphoproteins. J Virol 60:394–399
Nagy K, Clapham P, Cheingsong-Popov R, Weiss R (1983) Human T-cell leukemia virus type I: induction of syncytia formation and inhibition by patients' sera. Int J Cancer 32:321–328
Nakano S, Ando Y, Ichijo M, et al. (1984) Search for possible routes of vertical and horizontal transmission of adult T-cell leukemia virus. GANN 75:1044–1045
Okochi K, Sato H, Hinuma Y (1985) A retrospective study on transmission of adult T cell leukemia virus by blood transfusion; seroconversion in recipients. Vox Sang 46:245–253

Palker JT, Riggs E, Spragion D, Muir AJ Scearge R, et al. (1992) Mapping of homologous, amino-terminal neutralizing regions of human T-cell lymphotropic virus type I and II gp46 envelope glycoproteins. J Virol 66:5879–5889
Palker T (1990) Mapping of epitopes on human T-cell leukemia virus type I envelope glycoprotein. In: Blattner W (ed) Human retrovirology HTLV, Raven, NY, pp 435–445
Poiesz B, Rescetti F, Gazdar A (1980) Detection and isolation of type-C retrovirus particles from fresh and cultured lymphocytes of a patient with cutaneous T-cell lymphoma. Proc Natl Acad Sci USA 77:7415–7419
Poiesz BJ, Ruscetti FW, Reitz MS, Kalyanaraman VS, Gallo RC (1981) Isolation a new type-C retrovirus (HTLV) in primery uncultured cells of patient with Sezary T-cell leukemia and evidence for virus nucleic acids and antigens in fresh leukemic cells. Nature 294:268–271
Reddy EP, Mettus R, DeFreitas, E Wroblewska Z, Cisco M, et al. (1988) Molecular cloning of human T-cell lymphotropic virus type 1-like proviral genome from the peripheral lymphocyte DNA of a patient with chronic neurologic disorders. Proc Natl Acad Sci USA 85:3599–3603
Renjifo B, Chou K, Ramirez LS, Vallejo FG, Essex M (1996) Human T cell leukemia virus type I (HTLV-I) molecular genotypes and disease outcomes. J AIDS Hum Retrovirol 13:S146-S153
Rimsky L, Hauber J, Dukovich M, Malim MH, Langlois A, et al. (1988) Functional replacement of the HIV-1 rev protein by the HTLV-1 rex protein. Nature 335:738–740
Rudolph DL, Lal RB (1993) Discrimination of human T lymphotropic virus type-1 and type II infections by synthetic peptides comprising structural epitopes from the envelope glycoproteins. Clin Chem 39:288–292
Sagata N, Yasunaga T, Ikawa Y (1985) Two distinct polypeptides may betranslated from a single spliced mRNA of the genes of human T-cell leukemia and bovine leukemia viruses. FEBS Lett 192:37–42
Samuel K, Lautenberger J, Jorcyk C, Joseph S, Wong-Staal F, et al. (1984) Diagnostic potential for human malignancies of bacterially produced HTLV-I envelope protein. Science 226:1094–1097
Seiki M, Hattori S, Hirayama Y, Yoshida M (1983) Human adult T-cell leukemia virus: Complete nucleotide sequence of the provirus genome integrated in leukemia cell DNA. Proc Natl Acad Sci USA 80:3618–3622
Smith MR and Greene WC (1991) Molecular biology of the type I human T cell leukemia virus (HTLV-I) and adult-T cell leukemia. J Clin Invest 87:761–776
Suzuki T, Fujisawa JI, Toita M (1993a) The transactivator tax of human T cell leukemia virus type 1 (HTLV-1) interacts with cAMP responsive element (CRE) binding and CRE modulator proteins that bind to the 21 base pair enhancer of HTLV-1. Proc Natl Acad Sci USA 90:610–614
Suzuki T, Hirai H, Fujisawa J (1993b) A trans-activator tax of human T cell leukemia virus type 1 binds to NF-kappa B p50 and serum response factor (SRF) and associates with enhancer DNAs of the NF kappa B and CArG box. Oncogene 8:2391–2397
Suzuki T, Hirai H, Yoshida M (1994) Tax protein of HTLV-1 interacts with the rel himology domain of NF-kappaB/Rel involves phosphorylation and degradation of I kappa B alpha and RelA (p65)-mediated induction of the c-rel gene. Oncogene 9:3099–3105
Tajima K, Tominaga S, Suchi T, et al. (1982) Epidemiological analysis of the distribution of antibody to adult T-cell leukemia-virus-assocaited antigen: possible horizontal transmission of adult T-cell leukemia virus. Gann 73:893–901
Takahashi K, Takezaki T, Oki T (1991) Inhibitory effect of maternal antibody on mother-to-child transmission of human T lymphotropic virus type 1. Int J Cancer 49:673
Takatsuki K, Uchiyama T, Ueshima Y (1982) Adult T cell leukemia: proposal as a new disease and cytogenetic, phenotypic and functional studies of leukemic cells. Gann Monogr Cancer Res 28:13–22
Tanaka A, Takahashi C, Yamaoka S (1990) Oncogenic transformation by the tax gene of human T cell leukemia virus type 1 in vitro. Proc Natl Acad Sci USA 87:1071–1075
Tanaka Y, Tanaka R, Terada E (1994) Induction of antibody response that neutralize human T-cell leukemia virus type 1 infection in vitro and in vivo by peptide immunization. J Virol 68:6323–6231
Ugen K, Boyer J, Agadjanyan M, Wang B, Javadian A, et al. (1996a) Site-specific humoral immune responses generated in chimpanzees after DNA inocualtion with HIV-1 envelope expressing constructs. In: Brown F, Burton D, Mekalanos J, Norrby E (eds) Vaccines 96: Molecular approaches to the control of infectious diseases. Cold Spring Harbor, Cold Spring Harbor, NY, pp 61–65
Ugen KE, Agadjanyan M, Wang B, Kudchodkar S, Cho A, et al. (1995) Elicitation of immune responses in rodents after DNA inoculation with an HTLV-I envelop glycoprotein expressing construct: a model for vaccination efficacy. In: Brown F, Burton D, Mekalanos J, Norrby E (eds) Vaccines 95: Molecular approaches to the control of infectious diseases. Cold Spring Harbor, Cold Spring Harbor, NY, pp 129–134

Ugen KE, Fernandes L, Schumacher HR, Wang B, Weiner DB, et al. (1996b) Retroviruses and autoimmunity. In: H Freedman, R Noel, M Bendenelli (eds) Microorganisms and autoimmune diseases. Plenum, NY, pp 219–231

Ugen KE, Boyer JD, Wang B, Bagarazzi ML, Agadjanyan MG, et al. (1997) Nucleic acid immunization of chimpanzees as a prophylactic/immunotherapeutic vaccination model for HIV-1: prelude to a clinical trial. Vaccine 15:927–930

Ulmer JB, Donnelly J, Parker SE, Rhodes GH, Felgner PL, et al. (1993) Heterologous protection against influenza by injection of DNA encoding a viral protein. Science 259:1745–1749

Wang B, Boyer JD, Srikantan V, Ugen KE, Gilbert L, et al. (1995) Induction of humoral and cellular immune responses to the human immunodeficiency type 1 virus in non-human primates by in vivo DNA inoculation. Virology 211:102–112

Wang B, Ugen K, Srikantan V, Agadjanyan MG, Dang K, et al. (1993) Gene inoculation generates immune responses against HIV-I. Proc Nat Acad Sci USA 90:4156–4160

Weiss RA, Clapham P, Nagy K and Hoshino H (1985) Envelope properties of human T-cell leukemia viruses. Curr Top Microbiol Immunol 115:235–246

Yamanouchi K, Kinoshita K, Moriuchi R, et al. (1985) Oral transmission of human T-cell leukemia virus type-1 into marmoset (Callithrix jacchus) as an experimental model for milk-borne transmission. Jpn J Cancer Res 76:481–487

Yoshida M, Miyoshi I, Hinuma Y (1982) Isolation and characterization of retrovirus from cell lines of human adult T-cell leukemia and its implications in the disease. Proc Natl Acad Sci USA 79:2031

Yoshida M, Seiki M (1987) Recent advances in the molecular biology of HTLV-1: transactivation of viral and cellular genes. Ann Rev Immuno 15:541–559

Subject Index

Current Topics in Microbiology and Immunology

Volumes published since 1989 (and still available)

Vol. 186: **zur Hausen, Harald (Ed.):** Human Pathogenic Papillomaviruses. 1994. 37 figs. XIII, 274 pp. ISBN 3-540-57193-0

Vol. 187: **Rupprecht, Charles E.; Dietzschold, Bernhard; Koprowski, Hilary (Eds.):** Lyssaviruses. 1994. 50 figs. IX, 352 pp. ISBN 3-540-57194-9

Vol. 188: **Letvin, Norman L.; Desrosiers, Ronald C. (Eds.):** Simian Immunodeficiency Virus. 1994. 37 figs. X, 240 pp. ISBN 3-540-57274-0

Vol. 189: **Oldstone, Michael B. A. (Ed.):** Cytotoxic T-Lymphocytes in Human Viral and Malaria Infections. 1994. 37 figs. IX, 210 pp. ISBN 3-540-57259-7

Vol. 190: **Koprowski, Hilary; Lipkin, W. Ian (Eds.):** Borna Disease. 1995. 33 figs. IX, 134 pp. ISBN 3-540-57388-7

Vol. 191: **ter Meulen, Volker; Billeter, Martin A. (Eds.):** Measles Virus. 1995. 23 figs. IX, 196 pp. ISBN 3-540-57389-5

Vol. 192: **Dangl, Jeffrey L. (Ed.):** Bacterial Pathogenesis of Plants and Animals. 1994. 41 figs. IX, 343 pp. ISBN 3-540-57391-7

Vol. 193: **Chen, Irvin S. Y.; Koprowski, Hilary; Srinivasan, Alagarsamy; Vogt, Peter K. (Eds.):** Transacting Functions of Human Retroviruses. 1995. 49 figs. IX, 240 pp. ISBN 3-540-57901-X

Vol. 194: **Potter, Michael; Melchers, Fritz (Eds.):** Mechanisms in B-cell Neoplasia. 1995. 152 figs. XXV, 458 pp. ISBN 3-540-58447-1

Vol. 195: **Montecucco, Cesare (Ed.):** Clostridial Neurotoxins. 1995. 28 figs. XI., 278 pp. ISBN 3-540-58452-8

Vol. 196: **Koprowski, Hilary; Maeda, Hiroshi (Eds.):** The Role of Nitric Oxide in Physiology and Pathophysiology. 1995. 21 figs. IX, 90 pp. ISBN 3-540-58214-2

Vol. 197: **Meyer, Peter (Ed.):** Gene Silencing in Higher Plants and Related Phenomena in Other Eukaryotes. 1995. 17 figs. IX, 232 pp. ISBN 3-540-58236-3

Vol. 198: **Griffiths, Gillian M.; Tschopp, Jürg (Eds.):** Pathways for Cytolysis. 1995. 45 figs. IX, 224 pp. ISBN 3-540-58725-X

Vol. 199/I: **Doerfler, Walter; Böhm, Petra (Eds.):** The Molecular Repertoire of Adenoviruses I. 1995. 51 figs. XIII, 280 pp. ISBN 3-540-58828-0

Vol. 199/II: **Doerfler, Walter; Böhm, Petra (Eds.):** The Molecular Repertoire of Adenoviruses II. 1995. 36 figs. XIII, 278 pp. ISBN 3-540-58829-9

Vol. 199/III: **Doerfler, Walter; Böhm, Petra (Eds.):** The Molecular Repertoire of Adenoviruses III. 1995. 51 figs. XIII, 310 pp. ISBN 3-540-58987-2

Vol. 200: **Kroemer, Guido; Martinez-A., Carlos (Eds.):** Apoptosis in Immunology. 1995. 14 figs. XI, 242 pp. ISBN 3-540-58756-X

Vol. 201: **Kosco-Vilbois, Marie H. (Ed.):** An Antigen Depository of the Immune System: Follicular Dendritic Cells. 1995. 39 figs. IX, 209 pp. ISBN 3-540-59013-7

Vol. 202: **Oldstone, Michael B. A.; Vitković, Ljubiša (Eds.):** HIV and Dementia. 1995. 40 figs. XIII, 279 pp. ISBN 3-540-59117-6

Vol. 203: **Sarnow, Peter (Ed.):** Cap-Independent Translation. 1995. 31 figs. XI, 183 pp. ISBN 3-540-59121-4

Vol. 204: **Saedler, Heinz; Gierl, Alfons (Eds.):** Transposable Elements. 1995. 42 figs. IX, 234 pp. ISBN 3-540-59342-X

Vol. 205: **Littman, Dan R. (Ed.):** The CD4 Molecule. 1995. 29 figs. XIII, 182 pp. ISBN 3-540-59344-6

Vol. 206: **Chisari, Francis V.; Oldstone, Michael B. A. (Eds.):** Transgenic Models of Human Viral and Immunological Disease. 1995. 53 figs. XI, 345 pp. ISBN 3-540-59341-1

Vol. 207: **Prusiner, Stanley B. (Ed.):** Prions Prions Prions. 1995. 42 figs. VII, 163 pp. ISBN 3-540-59343-8

Vol. 208: **Farnham, Peggy J. (Ed.):** Transcriptional Control of Cell Growth. 1995. 17 figs. IX, 141 pp. ISBN 3-540-60113-9

Vol. 209: **Miller, Virginia L. (Ed.):** Bacterial Invasiveness. 1996. 16 figs. IX, 115 pp. ISBN 3-540-60065-5

Vol. 210: **Potter, Michael; Rose, Noel R. (Eds.):** Immunology of Silicones. 1996. 136 figs. XX, 430 pp. ISBN 3-540-60272-0

Vol. 211: **Wolff, Linda; Perkins, Archibald S. (Eds.):** Molecular Aspects of Myeloid Stem Cell Development. 1996. 98 figs. XIV, 298 pp. ISBN 3-540-60414-6

Vol. 212: **Vainio, Olli; Imhof, Beat A. (Eds.):** Immunology and Developmental Biology of the Chicken. 1996. 43 figs. IX, 281 pp. ISBN 3-540-60585-1

Vol. 213/I: **Günthert, Ursula; Birchmeier, Walter (Eds.):** Attempts to Understand Metastasis Formation I. 1996. 35 figs. XV, 293 pp. ISBN 3-540-60680-7

Vol. 213/II: **Günthert, Ursula; Birchmeier, Walter (Eds.):** Attempts to Understand Metastasis Formation II. 1996. 33 figs. XV, 288 pp. ISBN 3-540-60681-5

Vol. 213/III: **Günthert, Ursula; Schlag, Peter M.; Birchmeier, Walter (Eds.):** Attempts to Understand Metastasis Formation III. 1996. 14 figs. XV, 262 pp. ISBN 3-540-60682-3

Vol. 214: **Kräusslich, Hans-Georg (Ed.):** Morphogenesis and Maturation of Retroviruses. 1996. 34 figs. XI, 344 pp. ISBN 3-540-60928-8

Vol. 215: **Shinnick, Thomas M. (Ed.):** Tuberculosis. 1996. 46 figs. XI, 307 pp. ISBN 3-540-60985-7

Vol. 216: **Rietschel, Ernst Th.; Wagner, Hermann (Eds.):** Pathology of Septic Shock. 1996. 34 figs. X, 321 pp. ISBN 3-540-61026-X

Vol. 217: **Jessberger, Rolf; Lieber, Michael R. (Eds.):** Molecular Analysis of DNA Rearrangements in the Immune System. 1996. 43 figs. IX, 224 pp. ISBN 3-540-61037-5

Vol. 218: **Berns, Kenneth I.; Giraud, Catherine (Eds.):** Adeno-Associated Virus (AAV) Vectors in Gene Therapy. 1996. 38 figs. IX,173 pp. ISBN 3-540-61076-6

Vol. 219: **Gross, Uwe (Ed.):** Toxoplasma gondii. 1996. 31 figs. XI, 274 pp. ISBN 3-540-61300-5

Vol. 220: **Rauscher, Frank J. III; Vogt, Peter K. (Eds.):** Chromosomal Translocations and Oncogenic Transcription Factors. 1997. 28 figs. XI, 166 pp. ISBN 3-540-61402-8

Vol. 221: **Kastan, Michael B. (Ed.):** Genetic Instability and Tumorigenesis. 1997. 12 figs.VII, 180 pp. ISBN 3-540-61518-0

Vol. 222: **Olding, Lars B. (Ed.):** Reproductive Immunology. 1997. 17 figs. XII, 219 pp. ISBN 3-540-61888-0

Vol. 223: **Tracy, S.; Chapman, N. M.; Mahy, B. W. J. (Eds.):** The Coxsackie B Viruses. 1997. 37 figs. VIII, 336 pp. ISBN 3-540-62390-6

Vol. 224: **Potter, Michael; Melchers, Fritz (Eds.):** C-Myc in B-Cell Neoplasia. 1997. 94 figs. XII, 291 pp. ISBN 3-540-62892-4

Vol. 225: **Vogt, Peter K.; Mahan, Michael J. (Eds.):** Bacterial Infection: Close Encounters at the Host Pathogen Interface. 1998. 15 figs. IX, 169 pp. ISBN 3-540-63260-3

Printing: Saladruck, Berlin
Binding: Buchbinderei Lüderitz & Bauer, Berlin

MIX
Papier aus verantwortungsvollen Quellen
Paper from responsible sources
FSC® C105338

If you have any concerns about our products,
you can contact us on
ProductSafety@springernature.com

In case Publisher is established outside the EU,
the EU authorized representative is:
Springer Nature Customer Service Center GmbH
Europaplatz 3, 69115 Heidelberg, Germany

Printed by Libri Plureos GmbH
in Hamburg, Germany